"The authors' fascinating comparisons between Ghana and Cambodia tease out the complex and complicated relationship between agriculture commercialization, gender, and food security. Many of the chapters provide new insights and specific policy recommendations about the links between agricultural commercialization, food security, gender, and the right to food that could be applied across multiple countries."

Carolyn Sachs, *Professor Emerita of Rural Sociology, Pennsylvania State University, USA*

Agricultural Commercialization, Gender Equality and the Right to Food

This volume explores agricultural commercialization from a gender equality and right to food perspective.

Agricultural commercialization, involving not only the shift to selling crops and buying inputs but also the commodification of land and labour, has always been controversial. Strategies for commercialization have often reinforced and exacerbated inequalities, been blind to gender differences and given rise to violations of the human rights to food, land, work, and social security. While there is a body of evidence to trace these developments globally, impacts vary considerably in local contexts. This book systematically considers these dynamics in two countries, Cambodia and Ghana. Profoundly different in terms of their history and location, they provide the basis for fruitful comparisons because they both transitioned to democracy in the early 1990s, made agricultural development a priority, and adopted orthodox policies of commercialization to develop the sector. Chapters illustrate how commercialization processes are gendered, highlighting distinctive gender, ethnic, and class dynamics in rural Ghana and Cambodia, and the different outcomes these generate. They also show the ways in which food cultures are changing and the often-problematic impact of these changes on the safety and quality of food. Specific policies and legal norms are examined, with chapters addressing the development and implementation of frameworks on the right to food and land administration. Overall, the volume brings into relief multiple dimensions shaping the outcomes of processes of commercialization, including gender orders, food cultures, policy translation, national and sub-national policies, corporate investments and programmes, and formal and informal legal norms. In doing so, it not only offers insight on our case countries but also provides proposals to advance rights-based research on food security.

This book will be of great interest to students and scholars of food security, agricultural development and economics, gender, human rights, and sustainable development.

Joanna Bourke Martignoni is a senior researcher at the Geneva Academy of International Humanitarian Law and Human Rights, and an affiliate at the Graduate Institute's Gender Centre, Switzerland.

Christophe Gironde is a senior lecturer at the Graduate Institute of International and Development Studies (IHEID) in Geneva, Switzerland.

Christophe Golay is a senior research fellow and strategic adviser on economic, social, and cultural rights at the Geneva Academy of International Humanitarian Law and Human Rights, Switzerland.

Elisabeth Prügl is professor of international relations at the Graduate Institute, Geneva, Switzerland, where she heads the Institute's Gender Centre.

Dzodzi Tsikata is professor of development sociology and director of the Institute of African Studies (IAS) at the University of Ghana.

Earthscan Food and Agriculture

The Agricultural Dilemma
How Not to Feed the World
Glenn Davis Stone

Farming with the Environment
Thirty Years of Allerton Project Research
Chris Stoate

Net Zero, Food and Farming
Climate Change and the UK Agri-Food System
Neil Ward

Conservation Agriculture in India
A Paradigm Shift for Sustainable Production
Edited by A.R. Sharma

Agricultural Commercialization, Gender Equality and the Right to Food
Insights from Ghana and Cambodia
Edited by Joanna Bourke Martignoni, Christophe Gironde, Christophe Golay, Elisabeth Prügl, and Dzodzi Tsikata

The Sociology of Farming
Concepts and Methods
Jan Douwe van der Ploeg

Genetically Modified Crops and Food Security
Commercial, Ethical and Health Considerations
Edited by Jasmeet Kour, Vishal Sharma and Imtiyaz Khanday

For more information about this series, please visit: www.routledge.com/books/series/ECEFA/

Agricultural Commercialization, Gender Equality and the Right to Food

Insights from Ghana and Cambodia

Edited by
Joanna Bourke Martignoni, Christophe Gironde, Christophe Golay, Elisabeth Prügl, and Dzodzi Tsikata

Routledge
Taylor & Francis Group
LONDON AND NEW YORK
earthscan
from Routledge

First published 2023
by Routledge
4 Park Square, Milton Park, Abingdon, Oxon OX14 4RN

and by Routledge
605 Third Avenue, New York, NY 10158

Routledge is an imprint of the Taylor & Francis Group, an informa business

British Library Cataloguing-in-Publication Data
A catalogue record for this book is available from the British Library

Library of Congress Cataloging-in-Publication Data
A catalog record has been requested for this book

ISBN: 9781032063768 (hbk)
ISBN: 9781032063805 (pbk)
ISBN: 9781003202004 (ebk)

DOI: 10.4324/9781003202004

Typeset in Bembo
by codeMantra

Contents

Figures

Tables

Contributors

Daniel Adu Ankrah is a lecturer at the University of Ghana, Department of Agricultural Extension. Daniel has over 15 years' work experience in research, academia and development. His scholarship intersects agricultural policy, sustainable development, gender, agricultural innovation systems, climate variability and change, food security and rural advisory services.

Anna Antwi is a managing partner at GD Resources, a company committed to sustainable environmental processes for agriculture development. She has previously advised the Government of Ghana on a number of projects such as the Land Administration Project and the World Food Programme's Child and Maternal Nutrition Project. She was the right to food policy adviser at ActionAid Ghana, where she was involved in policy analysis, advocacy and campaigning. Dr Antwi was educated at Kwame Nkrumah University of Science and Technology (Bsc), University of the West Indies (MPhil) and University of Cape Coast (PhD).

Atudiwe P. Atupare is dean of the law faculty at the University of Cape Coast and a senior lecturer at the University of Ghana School of Law. Atudiwe holds a PhD from Queen's University in Canada. He teaches courses both at the undergraduate and graduate levels and has a wide-ranging experience and expertise in constitutional law, jurisprudence, criminal law, administrative law, employment and labour law and the law of torts.

Martha A. Awo obtained her master's in development studies at the Institute of Social Studies (ISS) in The Hague, The Netherlands, and her PhD at the Center for Research Development (ZEF) at the University of Bonn, Germany. She is currently working as a senior research fellow at the Institute of Statistical Social and Economic Research (ISSER), University of Ghana. As a former employee of the Ghanaian government's Ministry of Food and Agriculture, she has since 2000 been involved in implementing government projects and collaborated with international scientist to undertake studies in different areas of agriculture.

Alice Beban is a senior lecturer in sociology at Massey University, Aotearoa, New Zealand. She holds a PhD in development sociology from Cornell

University in the United States. Her research addresses land rights, agricultural production and gender concerns to understand people's changing relationships with land and water from a feminist political ecology perspective. Recent projects include research on 'land grabs' and redistributive land reform in Cambodia, and mapping cultural change in dam-affected communities along the Mekong river.

Joanna Bourke Martignoni is a senior researcher at the Geneva Academy of International Humanitarian Law and Human Rights and an affiliate at the Graduate Institute's Gender Centre. She is also the coordinator of the DEMETER project. Her research focuses on feminist legal theories and the role of international human rights norms in global governance. She has a particular interest in socio-legal approaches to economic, social and cultural rights, including the rights to food, land, education, decent work, sustainable development and social security.

Fred Mawunyo Dzanku is currently senior research fellow and head of the Survey and Statistics Division at the Institute of Statistical Social and Economic Research (ISSER), University of Ghana. Fred's area of expertise is applied microeconomic and econometric analysis, and he has more than 15 years of experience in conducting research in areas that include impact evaluations using experimental and quasi-experimental methods, ex ante evaluation of biotechnology and poverty reduction policy analysis. He holds a PhD in food and agricultural economics from the University of Reading, United Kingdom.

Promise Eweh is a PhD candidate in African studies at the University of Ghana, where he is currently conducting research on seed policy and food security. He is mainly interested in examining the theories and practices that are shaping Africa's development, and their impacts on common people, with particular attention to agriculture, employment and the informal economy. He looks forward to a future of engagement and collaborative research with local communities and persons of diverse backgrounds interested in the living conditions of Africa's working people.

Christophe Gironde is a political economist, currently working as a senior lecturer at the Graduate Institute of International and Development Studies (IHEID) in Geneva. His main interests are agrarian change and human development. He has extensive field research experience in rural Vietnam and Cambodia, and previously in the Republic of Congo and Burundi. Before joining the Institute in 2004, he worked as a researcher in Norway (Fafo, International Studies) and lived in Vietnam (1996–97) in the frame of his doctoral thesis. He is currently working on land commercialization processes and their consequences for rural livelihoods, gender relations and food security.

Christophe Golay is a senior research fellow and strategic adviser on economic, social and cultural rights at the Geneva Academy of International

Humanitarian Law and Human Rights. He is visiting professor at the Graduate Institute, Geneva. From 2001 to 2008, he was legal adviser to the first UN Special Rapporteur on the right to food. Since 2009, he has provided legal advice in relation to the negotiation, adoption and implementation of the UN Declaration on the Rights of Peasants and Other People Working in Rural Areas. He has published several books and articles on the promotion of the right to food and is a co-applicant for the DEMETER project.

Saba Joshi is a lecturer (assistant professor) in gender and development at the Department of Politics at the University of York. She was a research assistant and a PhD student in the DEMETER project between 2015 and 2019. Her doctoral thesis examined the politics of land grabbing in Cambodia, drawing on extensive field-based research. In 2020–2021, she was Swiss National Science Foundation Research Fellow based at the University of Oxford, where she researched women's activism over dispossession in contemporary India. Her research and teaching interests lie in the intersecting fields of feminist political economy, agrarian politics and authoritarian rule.

Muy Seo Ngouv is a researcher and lecturer at the Center for the Study of Humanitarian Law at the Royal University of Law and Economics in Cambodia. She has conducted legal research and training on various topics relating to human rights law and humanitarian law. Her main research interests include the areas of gender and the rights to food and land, gender in business and human rights, women's leadership in legal professions, fair trial rights, women's rights and children's rights.

Amaury Peeters is director of Research and Development at Louvain Cooperation in Belgium. Prior to taking up that position, he spent six years as the Cambodia director for Louvain Cooperation, and in that capacity, he contributed to a number of research for agricultural development projects focusing on food and nutrition.

Elisabeth Prügl is professor of international relations at the Geneva Graduate Institute. Her scholarship focuses on gender and international governance, and she is the responsible applicant for the DEMETER project. She has published extensively on issues ranging from the regulation of atypical forms of women's labour (including in agriculture) to gender expertise in international governance. In 2019, she received the Eminent Scholar Award of the Feminist Theory and Gender Studies section of the International Studies Association (ISA).

Fenneke Reysoo, is an anthropologist, gender expert, and emeritus professor of the Geneva Graduate Institute. In addition to her work on Cambodia, she has carried out research and published on gender relations in Bangladesh, Burkina Faso, France, Mali, Mexico, Morocco, Sri Lanka and Switzerland.

Kim Thida is a psychologist in private practice and a PhD candidate in social psychology at the University of Hong Kong. She spent nearly three years as the research coordinator for Louvain Cooperation in Cambodia and also worked as a researcher on several agricultural development projects in Cambodia for the German Development Cooperation agency (GIZ). She holds master's degrees in psychology and in gender and development studies from universities in the Philippines and in Thailand.

Andres Torrico Ramirez is a researcher, specialized in rural development and indigenous communities. He holds a master's in development studies from the Graduate Institute of International and Development Studies and a master's in public policy and management from the Catholic University in Bolivia. He is research associate at the Graduate Institute of International and Development Studies in Geneva in the frame of the DEMETER project. Prior to joining the institute, he worked for the Research Center on Promotion of Peasantry in Bolivia (CIPCA).

Gertrude Dzifa Torvikey is a research fellow at the Institute of Statistical, Social and Economic Research (ISSER), University of Ghana. She also works with *Feminist Africa* journal at the same university. Her research interests are in agrarian livelihoods, migration, labour and gender issues. She has worked on agricultural projects such as DEMETER – Gender, Land and the Right to Food, COVID-19 and Africa's Food Systems, Land and Agricultural Commercialisation in Africa (LACA) and Agricultural Policy Research in Africa (APRA), among others.

Dzodzi Tsikata is professor of development sociology and director of the Institute of African Studies (IAS) at the University of Ghana. Her scholarship is in the areas of gender and development policies and practices; the gendered political economy of agrarian change; and informal labour relations and conditions of work and she has published extensively on these issues. Her publications include the co-edited book (with Ruth Hall and Ian Scoones) *Africa's Land Rush: Implications for Rural Livelihoods and Agrarian Change* (Boydell and Brewer Ltd; 2015). She is a co-applicant for the DEMETER project.

Acknowledgements

Research leading to this publication has been supported by a grant from the Swiss Programme for Research on Global Issues for Development co-funded by the Swiss Agency for Development and Cooperation and the Swiss National Science Foundation, Grant Number 400540.

Agricultural commercialization, gender equality, and the right to food

Joanna Bourke Martignoni, Christophe Gironde, Christophe Golay, Elisabeth Prügl, Fenneke Reysoo, and Dzodzi Tsikata

The United Nations Agenda for Sustainable Development reaffirms the long-standing global goal of ending hunger and seeks to accomplish this by 2030. But the challenge is massive, and according to the UN's 2021 report on *The State of Food Security and Nutrition in the World*, the goal is unlikely to be achieved. Almost 10% of the world population faced hunger in 2020 and more than 30% were unable to access adequate food. Moreover, food insecurity has massively increased in the shadow of the COVID-19 pandemic (FAO, IFAD, UNICEF, WFP and WHO, 2021).

The UN agencies' report not only identifies conflict, climate change, and economic downturns as drivers of food insecurity but also emphasizes the key roles of poverty and inequality as underlying structural factors that make some people more prone to experiencing hunger than others. Indeed, there is today a widespread consensus that food security is not a matter of insufficient availability of food but more importantly of people not having access to land to produce food for their own consumption, or income from work or social security entitlements to be able to purchase food. Although food insecurity is thus recognized as a matter of distribution, international development policies continue to focus on increasing the production of food and non-food commodities. They seek to boost food supplies and farmers' incomes through the commercialization of agriculture, counting on private sector investments to enhance productivity and the free market to distribute food efficiently. The theory is that commercialization will increase both the availability and accessibility of food. In the long run, efficient markets are said to select the most productive farmers and insert them into global value chains, while farmers who are unable to compete move into other livelihoods. In this paradigm, inequality often appears as an afterthought in the trade-off between equality and efficiency, an unfortunate side effect that other policies must seek to address. Yet, if food insecurity is a matter of accessibility, inequality is part of the problem, which neoliberal development policies may aggravate rather than alleviate. There is ample literature to show that unfettered markets tend to generate inequalities along multiple axes of difference (Losch, 2004). These include ethnicity and indigeneity; they also include gender. In

DOI: 10.4324/9781003202004-1

commercialized agriculture, the unpaid farm and reproductive labour disproportionately performed by women are simultaneously central to and invisible within rural economies. The so-called 'triple burden' of farming work, off-farm labour, and reproductive labour that women provide within families and communities is symptomatic of the 'super-exploitation of women and the environment' (Federici, 2019) and ultimately, neoliberal capitalism's 'crisis of care' (Fraser, 2016).

The contemporary food system is increasingly global. International trade in agricultural products has expanded rapidly since the 1980s, entailing the development of global value chains that link producers and consumers across large distances. These value chains are reliant on a highly concentrated private sector for the organization of inputs, production, trade, processing, and retailing. Indeed, corporations increasingly are not only involved in the governance of international supply chains but also encouraged to partner with public actors to foster commercialization as a means of development and to participate in setting standards (Clapp and Fuchs, 2009). Not surprisingly, the global food regime celebrates technical solutions to increase food production and reduce costs, increasingly adopting biotechnological innovations, such as genetically modified organisms and automated field operations (Hopma and Woods, 2014; Bourke Martignoni, 2021).

Since the mid-2000s, there has been an inexorable rise in food prices, driven in part by population growth, increased living standards, and consequently food consumption, in particular in emerging economies with huge populations such as China (McMichael, 2020), and in part by various non-food industries, such as biofuels, as environmental policies have supported the production of these crops as part of the transition away from fossil fuel energies (Dufey et al., 2007). The rising demand for agricultural products has spawned a wave of large-scale land acquisitions and the territorial expansion of industrial farms, in particular in countries in the Global South. High-income countries have acquired large tracts of land around the world to secure their own provision of food and animal feed products, commodities for biofuels, and raw materials for industry (Meyers and Meyer, 2008; Matondi et al., 2011). While this global land rush has been described as 'unprecedented' (White et al., 2012), there is doubt as to whether it constitutes a 'distinct historical phenomenon' (Margulis et al., 2013), a 'turning point' (Oya, 2013, p. 1548) in agrarian change in that land has now become a financial asset (Smaller and Mann, 2009), or instead a continuation of processes of land expropriation that reach back to colonial periods (Borras et al., 2011, p. 212). Regardless, a significant territorial expansion and intensification of commercial agriculture can be observed in Cambodia, where large-scale, mono-crop plantation agriculture has exploded (Carney, 2021), and in Ghana, where small landholding producers have been induced to develop cash crops on their own land or through 'value-chain agriculture' (Hall, 2011; McMichael, 2013), including via contract farming and out-grower schemes (Hall et al., 2017).

These developments have not gone without criticism. Growing alarm over the financialization of global agriculture, price fluctuations due to speculation, and interruptions of food supplies during COVID-19 lockdowns have led to increased demands for re-localizing food production (Clapp and Helleiner, 2012; Suarez Franco, 2021). Long before the pandemic, a variety of social movements, the most visible of which has been La Via Campesina (LVC), called for an end to the globally integrated and corporate-dominated food system and for food to be produced through ecologically sound and sustainable methods (La Via Campesina et al., 2007; McKeon, 2014). They have demanded 'food sovereignty', that is the right of peoples to healthy and culturally appropriate food and to define their own food and agriculture systems. They also have denounced the human rights violations that have accompanied land grabbing, the violence used to evict peasants from their lands and the economic, social, and environmental costs of unregulated neoliberal development. In parallel, feminist activists and scholars have highlighted the gender-unequal outcomes of neoliberal policies. By taking disembodied 'rational' actors as their starting points, these policies ignore women's reproductive labour and end up promoting rural development at their expense, often not recognizing women as farmers in their own right and side-lining them into unpaid labour (Gladwin, 1991; Whitehead, 2008; Razavi, 2009; Tsikata and Amanor-Wilks, 2009). These critiques have found resonance in international human rights fora, in particular with the UN Special Rapporteur on the Right to Food, the UN Committee on the Elimination of Discrimination Against Women (CEDAW), and in the intergovernmental Committee on World Food Security (CFS). They have accelerated the adoption of international norms on the right to food and the rights of peasants that incorporate provisions on gender equality and the rights of women in rural areas (Bourke Martignoni and Claeys, 2022). They also have led to the formulation of voluntary guidelines, by both the CFS and the World Bank, that attempt to temper the most exploitative features of land and agricultural commercialization and alleviate their discriminatory effects.

The research in this book takes the right to adequate food as its normative starting point to evaluate the unequal impacts of neoliberal agricultural policies and laws on the livelihoods of rural populations and their food security. In parallel, we explore the implementation of policies and laws that seek to advance gender equality and right to food. Propelled by Margulis' (2013) argument that food security is governed by an international 'regime complex' that combines neoliberal ideas with rights-based understandings, we examine how international, national, and subnational policies and laws navigate the tensions between these sometimes-contradictory approaches. Our findings are based on research in Cambodia and Ghana. Profoundly different in terms of their histories, cultures, and geographical location, the two countries provide the basis for fruitful comparisons because they have made agricultural development a priority and adopted neoliberal approaches to accelerate agricultural and land commercialization. As a result, they have undergone rapid agrarian transformations and are experiencing considerable commercial pressure on land. Each is also a party to international human rights treaties that create obligations to realize gender equality, the right to adequate food, and the inter-related rights to decent work and social security. In studying how the two countries implement these policies and the outcomes this has generated in their diverse contexts, we seek to discern distinctive mechanisms and patterns while also contributing to a better understanding of how to advance gender equality and the achievement of the right to food globally.

In the next section of this introduction, we provide an overview of international policies on food security, the right to food, and gender equality, highlighting the tensions between neoliberal and rights-based approaches. Next, we review existing literature that explores how the commercialization of land and agriculture relates to food security and how processes of commercialization are gendered. We then introduce the DEMETER project, under which the research for the chapters in this collection was carried out, present the research questions, discuss our key concepts, and explain our research design and methods. Finally, we provide an overview of the chapters that draw out the broad questions they raise and their findings.

Governing food security: neoliberalism, the right to food, and gender equality

In the post–World War II era, food security policies were typically understood as policies to modernize agriculture. The experience of food shortages during the war paired with Malthusian anxieties over population growth fostered a productivist stance towards agriculture. Food security was considered a matter of national security; that is, the goal was to ensure reliable food supplies (in the Global North) and stave off famine and malnutrition as well as rural to urban migration (in the Global South). This understanding shifted in the 1980s, when an influential World Bank Report (World Bank, 1986) introduced the distinction between chronic and transitory food insecurity, suggesting that food security was a matter of purchasing power as much as sufficient supplies. In addition, Amartya Sen's influential book *Poverty and Famine* (1981) reinterpreted food security from the perspective of households and individuals, linking it to issues of inequality, entitlement, and freedom. Sen's ideas were also incorporated into international policy discourses, including the UN Development Programme's (UNDP) first Human Development Report in 1990. Food security thus increasingly became associated with equal access to productive resources and income through work or from social protection schemes. The definition of food security accepted at the 1996 World Food Summit reflects this broadened understanding: 'Food security exists when all people, at all times, have physical and economic access to sufficient, safe and nutritious food that meets their dietary needs and food preferences for an active and healthy life' (World Food Summit: Plan of Action, 1996). The core elements of this definition are often summarized as availability, access, stability, and utilization, with the latter including the adequacy of diets, nutritional well-being, and non-food inputs, such as clean water (FAO, 2006).

Yet, commercialization remained a core part of these policies, designed to ensure that food production kept up with projections (such as those put forward at the 2008 World Food Summit) that the world population would reach 9 billion by 2050 (Maye and Kirwan, 2013, p. 1) and the assumption that it would enable rural populations to increase their incomes. In the 1980s, neoliberal prescriptions were firmly encoded in the global food regime. The Uruguay Round of trade negotiations, launched in 1986, focused on the integration of agriculture within trade and investment regimes that favour liberalized market mechanisms. In parallel, structural adjustment programmes imposed policies on indebted countries in the South that dismantled national government control over food and agriculture. National land titling systems were developed – frequently with the support of international financial institutions and bilateral aid programmes – to facilitate the creation of land and agricultural commodities markets (De Schutter, 2015). Financial incentives were made available to prioritize economies of scale and promote large-scale, capital-intensive agriculture, while support for smallholder farmers was reduced. Foreign direct investment in the agricultural and land sectors in the South gained momentum (Ashwood et al., 2020).

At the same time, the imposition of neoliberal agricultural policies has long been contested, pitting the interests of countries with large-scale agriculture against those dominated by smallholders, and those of the South against the North. Today, the process of liberalizing trade in agriculture has stalled against the resistance of emerging economies, such as India and Brazil. Developmental priorities, in addition to social protection, climate change adaptation, and mitigation, have become major considerations in the future shape of global food systems (Saab, 2019). The idea of the right to food has gained traction within international food policy forums. It is also being promoted through the activism of transnational social movements who view the realization of the right to food as an essential component of food sovereignty frameworks (Bourke Martignoni, 2020). The Universal Declaration of Human Rights (article 25), the International Covenant on Economic, Social and Cultural Rights (article 11), and many regional human rights instruments and national constitutions contain provisions on the human right to adequate food (Golay, 2011; Ziegler et al., 2011; Bourke Martignoni, 2020). In its interpretive General Comment No. 12, the UN Committee on Economic, Social and Cultural Rights notes that the right to adequate food means the right for all people to have 'physical and economic access at all times to adequate food or means for its procurement' (CESCR, 1999, paragraph 6). The main benefit of a right to food-based approach is its emphasis on the responsibilities of states and other duty bearers to respect and protect the right without discrimination, which also means that they must remove any structural barriers to universal access to food (CESCR, 1999; Ziegler et al., 2011; De

Schutter, 2014; Fakhri, 2020). At the international level, states, international organizations and civil society actors recognize that the right to food plays an integral role in the achievement of food security, and legislation and policies for the implementation of the right to adequate food have been adopted in several countries (Graziano Da Silva et al., 2011; Monsalve Suárez, 2013, Graziano Da Silva, 2019).

The right to food approach resonates with efforts to integrate gender equality goals into the international governance of food security. Rural women and gender equality figure heavily in discourses on the right to food and sustainable development. Various instruments, including the Universal Declaration of Human Rights (article 2), the International Covenant on Economic, Social and Cultural Rights (articles 2.2 and 3), and the Convention on the Elimination of All Forms of Discrimination against Women (article 14) acknowledge the linkages between women's equal rights and the realization of the right to food (Bourke Martignoni, 2018). The outcome document of the 1995 World Conference on Women in Beijing called for the systematic mainstreaming of gender considerations into all policies, and the UN specialized agencies in Rome (FAO, WFP, IFAD) and the World Bank have established gender strategies within their planning cycles. The UN's 2030 agenda for sustainable development recognizes that achieving the goal of eliminating hunger requires attention to the important role women play in food production through small-scale farming. And in 2018, the UN Commission on the Status of Women (CSW) reaffirmed the right to food and recognized 'the crucial contributions of rural women to local and national economies and to food production and to achieving food security and improved nutrition' (CSW, 2018, para. 17).

The implementation of gender mainstreaming within food and agricultural commercialization policies and legislation has also attracted criticism. For example, giving women individual land titles, which in neoliberal policy circles are considered a prerequisite for creating land markets, is often promoted as a form of empowerment. In contexts of unequal gender relations, however, individual joint titling initiatives have actually led to the dispossession of women and to the alienation of communal lands that may have previously been used to supply food (Deere and León, 1987; Agarwal, 1994; Lastarria-Cornhiel, 1997; Razavi, 2003, 2007; Tsikata, 2009; Levien, 2017; Bourke Martignoni, 2018). More broadly, the 'smart economics' approach of the World Bank and other international financial institutions tends to subordinate gender equality to gains in economic efficiency and instrumentalizes women as vectors of economic growth, household food security, and broader development outcomes without changing intrinsic structural inequalities (Roberts and Soederberg, 2012; Calkin, 2015; Prügl, 2016; Esquivel, 2017). In the regime complex on food security, neoliberal ideas have thus become tightly interwoven with the language of rights and ideas of gender equality in particular. The suggestion that gender equality is both a matter of human rights and economic efficiency has become an international mantra (Prügl and Joshi, 2021).

The gendered impacts of commercialization on food security

An extensive body of empirical literature has developed to assess the diverse impacts of agricultural commercialization and export-led agriculture, including with regard to gender equality. Debates in this literature centre on the productive efficiency of small farms compared to large farms, the marginalization and disappearance of small farmers, the impacts on food security, and gendered effects with regard to labour and access to resources.

A core argument has been that large-size commercial farms, due to their capacity to mobilize capital and technology, can be the drivers of a productive modernization that small-scale farms cannot achieve (Collier, 2008). Scholars suggest that capital-intensive farms have greater capacity to adopt new crops, adapt to changing demand, generate higher yields and therefore have better access to global markets and value chains than most small landholders (Byres, 2012). In the long run, it is argued, the development of large farms will benefit smaller farms through technology transfer and by facilitating access to new markets. In this scenario, it is posited that market efficiency will increase the availability and accessibility of food as successful farmers can insert themselves into global value chains while those who are less competitive will move to non-farming livelihoods where they can earn income to purchase food.

But the comparative advantage argument about large-scale farming has been controversial for decades, as small-scale farmers also have commercialized successfully (Ellis and Biggs, 2001). In the case of Southeast Asia, the participation of smallholders in commercial agriculture has been widespread (Hall, 2011), and there are many contexts in which they have been successful in growing cash crops (Delarue, 2011; Sikor, 2012). Their performance depends in particular on the support they receive from governments (Gouyon, 1995; Fox and Castella, 2013). The technical superiority of large-scale farming in the generation of agricultural surpluses has also been questioned in the case of sub-Saharan Africa (see Baglioni and Gibbon, 2013).

Contrary to the trickle-down paradigm and its associated 'win-win' rhetoric, a majority of empirical studies have concluded that agricultural commercialization, including large-scale land acquisitions, have been detrimental to most small landholders. First, smallholders lose farming land and free access to natural resources such as wild food, grazing, and fishing areas. In many cases, if not systematically, losses may be aggravated due to the absence of formal land rights, poor regulatory enforcement, and a lack of access to independent legal mechanisms to adjudicate land rights in cases of dispossession. Second, small landholders may not have the resources needed to invest in commercial crops. Third, they may not be able to make enough profits to guarantee a decent standard of living given volatile global value-chain prices and the fact that they are competing against large-size farmers (van der Ploeg, 2008; Spieldoch and Murphy, 2009; De Schutter, 2011; Amanor, 2012; Borras and Franco, 2012). Finally, the impacts of environmental degradation and climate change are increasingly making themselves felt in agricultural production and small-scale farmers often have less ability to invest in climate adaptation (Saab, 2019).

Effects on *food security* have also been problematic, as, in the words of Bello (2008), liberalization has amounted to the 'manufacture of a global food crisis'. A substantial share of land in the current wave of acquisitions has been used for the production of non-edible crops, fodder, and/or food crops that are exported (Daniel, 2011; Cotula et al., 2014; Rulli and D'Odorico, 2014; Sans and Combris, 2015; Titche, 2017). In the liberalization paradigm, this is not considered a problem because the idea is that food security in rural areas may be achieved through the importation of food. But this idea has been the subject of much debate, and there are recommendations that additional policy measures be taken to 'alleviate many of the possible adverse transitional consequences' (Pingali and Rosegrant, 1995, p. 184). Indeed, emerging evidence, including from the DEMETER project, finds little improvement in nutritional outcomes as a result of commercialization (Carletto et al., 2017). In our case regions in Cambodia many could not afford to purchase enough and quality food, and borrowing has become crucial, not only for productive activities, but also to buy enough to eat (Gironde et al., this volume). And our Ghana study identifies commercialization-induced distress sales of property to combat hunger (Dzanku et al., this volume), raising questions about the impacts of commercialization on marginalized social groups.

These impacts have been deeply gendered. Over several decades, feminist political economists have shown that with commercialization and associated class differentiation, 'women's independent farming came under increasing pressure, while many men were able to solidify their command over land, labour, and capital resources' (Razavi, 2009, p. 203). Gender differentiated access to and control over resources of all kinds have reproduced gender inequalities since the colonial period (Agarwal, 1994; Lastarria-Cornhiel, 1997; Deere and Leon, 2001; Razavi, 2003; Tsikata, 2016). Intra-household relations of production and reproduction are a crucial transmission belt for the gendered impacts of commercialization (Elson, 1998; Naidu and Ossome, 2016). As men's subsistence labour becomes wage labour or is incorporated into value chains, this has entailed the 'productive deprivation' of women: their labour gets displaced and new gender divisions of labour emerge (Harriss-White, 2005). Women's labour remains unremunerated, loses value, and becomes readily available as a low-cost input into projects of commercialization. Thus, studies of export-oriented agriculture have shown that investment schemes typically have created a low-paid, often seasonal, and disproportionately female labour force, although differentiated by other status positions (Barrientos et al., 1999; Barndt, 2008; Bigler et al., 2017; Sulle and Dancer, 2020). Moreover, gendered inequalities, in intersection with other status positions, have facilitated the extraction of surplus value from agriculture, establishing 'chains of exploitation' with rural women at the bottom and agribusiness companies at the top (Maffii, 2009; Luna, 2019). Research has concluded that a combination of pre-existing gender inequalities, women's lack of power to claim rights, and the gender blindness and biases of projects are responsible for the poor livelihood outcomes and gendered impacts of commercial agriculture projects (Izumi, 2007; Mutopo, 2011; Behrman et al., 2012; Daley and Park, 2012).

Based on this scholarship, studies of the new wave of land grabbing in the early 21st century confirmed that pre-existing gender inequalities shaped the outcomes for the dispossessed, creating particularly onerous structural barriers for women to access new opportunities (Izumi, 2007; Chu, 2011; Behrman et al., 2012; Julia and White, 2012; Daley and Pallas, 2014; Doss et al., 2014; Mutopo and Chiweshe, 2014; Tsikata and Yaro, 2014; Park and White, 2017). Employment opportunities created by agricultural investments were insecure and mainly reserved for men, and few women were able to take advantage of out-grower contracts. Instead, women lost the resources they derived in the past from harvesting the commons (Brandt Brocgaard et al., 2017). While enjoying ancillary benefits provided through corporate social responsibility (CSR) programmes, these were mainly targeted at their domestic activities and at their children, providing a meagre substitute for what was lost (Behrman et al., 2012; Julia and White, 2012; Piacenza, 2012; Daley and Pallas, 2014; Tsikata and Yaro, 2014).

In an influential article in 2011, Tania Murray Li (2011) argued that the impacts of land commercialization should be assessed through the lens of changes in labour relations. Indeed, various studies reported that one effect of large-scale land acquisitions was an increase in women's workloads. In Cambodia for example, the agricultural wage work in areas impacted by economic land concessions (ELC) was associated with more reproductive work for women, generating time poverty with frequently negative consequences for food security (Ironside, 2009; Maffii, 2009; Paramita, 2013; Beban and Bourke Martignoni, 2021; Gironde et al., 2021). In upland rice systems, when farmers reduced fallows due to pressure to raise production, weeds frequently increased, and gender roles were re-inscribed as older women performed more weeding or applied herbicides, while men and younger women migrated (Pierce Colfer, 2013). Similar effects were reported from Ethiopia, where large-scale land transactions resulted in increased labour time for women (Hajjar et al., 2019). More broadly, in his historically and geographically wide-ranging comparative review of the effects of land dispossession on women, Levien (2017) illustrates how changes in the gender division of labour linked to land commercialization were invariably disadvantageous to women, though this differed by class and caste.

Critical engagement with the land-grabbing literature resulted in a second generation of research that pays attention to longer-term processes of land and agrarian commercialization (Edelman et al., 2013; Oya, 2013). Studies in this vein have broadened the discussion from the dispossession of smallholder farmers to examining the processes that lock them into global agricultural value chains as plantation workers, contract farmers, and small- and medium-scale farmers producing for export markets. Lending nuance to processes of commercialization, this research shows that outcomes differ depending on local contexts and investment practices (Gironde et al., 2015). Commercialization seems to increase household incomes in the aggregate, but also generates considerable inequality as it leverages patriarchal norms and political economies in ways that tend to favour men (Gironde et al., this volume; Hall et al., 2017; Park and Maffi, 2017; Park and White, 2017).

The literature thus casts doubt on commercialization as a pathway to achieving global food security. Food production and earning off-farm income both emerge as problematic for women, producing land loss and seasonal food shortages on the one hand and changing divisions of labour, low wages, and new dependencies on men's income on the other. While contexts and intersectional inequalities matter, unfettered land and agricultural commercialization seems to disproportionately benefit those with access to power and resources. Understanding how such power relations affect household food security requires a closer look at situated contexts.

Approach and methodology

The chapters in this book emerged from an international research project on land commercialization, gendered agrarian transformation, and the right to food – the DEMETER project. Started in 2015 by a research partnership of scholars from Cambodia, Ghana, and Switzerland, the project asked three questions: (1) how does the commercialization of land and agriculture impact food security? (2) How do local, national, and international norms and policies shape these outcomes? and (3) how do processes of commercialization and associated policies, laws, and practices affect gender equality and the right to food?

To address these questions, as an interdisciplinary team of scholars, we combined critical approaches from political economy, gender studies, and socio-legal studies, and applied these within a human rights-based approach to development. This led us to conceptualize commercialization as a type of agrarian change, to think of gender in relational terms, and to use it as an analytical lens.

Commercialization as agrarian change

The development of commercial agriculture and the relative decline in subsistence agriculture is a universal phenomenon (Friedmann and McMichael, 1989; Mazoyer and Roudart, 1997). It involves the replacement of staple food crops that were used mainly to feed their producers with crops that are farmed for sale. However, 'agricultural commercialization means more than the marketing of outputs' (Pingali and Rosegrant, 1995, p. 171). It reorganizes the entire process of agricultural production, changing the character of land, labour, capital, inputs, and technology, which increasingly become commodities to be purchased, rented, and sold (Li, 2011). Commercial transactions, in particular loans and repayments, become the primary mechanisms through which rights over productive resources are granted or rescinded. Personal- and community-based relationships, rights, obligations, and ethical principles such as redistribution, reciprocity, and solidarity (as identified by Polanyi, 1944) become less relevant in this new, commercial environment. Agricultural commercialization can thus be conceptualized as a process through which agricultural land, labour, and capital are increasingly distributed and allocated through monetary transactions, according to market mechanisms and commodity pricing structures.

Agricultural commercialization is commonly associated with the consolidation of rural land holdings, increased capital investment and indebtedness, the widespread use of agro-industrial technologies, and the employment of wage labour. In order to produce commercially, smallholder farmers frequently find themselves under pressure to invest in agricultural inputs and machinery, with the growing need for capital and associated indebtedness driving forced sales of commodities and land. This tends to promote the dominance of large-scale, agro-industrial farming businesses. For small- and medium-scale farmers, it often means their transformation into outgrowers in subcontracting arrangements for agro-companies, or they may become independent commercial farmers, producing mainly for the market and selling crops through brokers. Moreover, beyond agriculture, commercialization is usually accompanied by the development of non-agricultural livelihood activities, wage labour, and migration. Land is increasingly used for purposes other than the production of agricultural commodities, including conservation, special economic zones and infrastructures, mining, tourism, and residential housing (Zoomers, 2010). Thus, the growing importance of commercial agriculture is part of a broader process of agrarian change.

Gender as a relation and analytical lens

Following a political economy approach, we conceptualize gender as a social relation that is constitutive of divisions of labour and reproduces hierarchies in intersection with other status distinctions. The gender division of labour has long been theorized as generating a division between production and reproduction, between women's work and men's work, a division that underwent fundamental change during industrialization to create the figure of the unproductive housewife, which has been generalized to countries beyond Europe (Mies, 2012). But the housewife is not a universal figure. Feminist political economists have brought into view the changing character of social reproduction in contexts of commercialization, historizing women's work. In so doing, they have shifted the narrow focus of political economy from surplus production to matters of care and the production of life (Razavi, 2009; Fraser, 2016; Bhattacharya, 2017). In this book, we focus on these processes, highlighting in particular matters of food and nutrition as critical aspects of social reproduction.

But we approach gender not only as a social relation that can be historically described. Gender also works by signifying power, assigning status hierarchies, and in this way regulating access to resources (Scott, 1986). It does so in intersection with other status positions, such as ethnicity or indigeneity (Menon, 2015). In the context of this project, we thus employ gender also as a lens, a way of looking at commercialization that makes visible the changes in power relations it brings about. It is a lens that refracts other axes of difference and, in this way, trains our focus on the reproduction of multiple inequalities. Gender thus articulates with the concept of rights: it makes visible injustices and offers the grounds for an assessment of outcomes.

Research design and cases

Our research was designed to enable comparisons between countries and between regions within countries that exhibit different patterns of agricultural commercialization. We selected two countries, Cambodia and Ghana, for in-depth case studies. For Cambodia, the shift towards economic liberalization coincided with the end of the civil war in the 1990s. The 1993 elections under UN supervision installed a government intent on moving the country to a free-market economy while strengthening its control over land and natural resources to consolidate its political power. In the 1990s and early 2000s, development institutions such as the World Bank supported large-scale land mapping and registration policies that set the scene for agricultural and land commercialization (Biddulph, 2010; Diepart and Sem, 2018). A new system of ELCs was established, leveraging neo-patrimonial political structures that empowered elites while generating new inequalities in the countryside. It attracted domestic and foreign investments that resulted in massive commercial logging and land consolidation along with the internal migration of people in search of farming land. It also led to the dispossession of smallholder and indigenous farmers and to diminished access to forests, communal grazing lands, and fisheries. In practice, the process of agricultural and land commercialization in Cambodia has been conflictual and marked by the violent grabbing of land from smallholders, dependence on volatile boom crops such as cassava and rubber, widespread indebtedness, environmental degradation, and increased inequalities within many communities and households.

In Ghana, the return to democracy in 1993, after years of populism and military rule, did not entail a major shift in economic policies. Rather, the transformed Rawlings government remained true to the neoliberal economic logic that had informed its structural adjustment policies during the 1980s. With a strong export-oriented sector since colonial times focused on the production of cocoa, Ghana's agricultural production has long been deeply integrated into global markets. Since the 1990s, national development policies in the food and agricultural sector have prioritized strengthening investment in agro-industrial companies as key engines of growth, and these policies have garnered extensive support from international development actors. A multi-year, internationally funded land administration project (LAP) was put in place starting in 2011, geared towards establishing a land governance framework through the promotion of titling and registration in an effort to provide security of tenure and, in turn, to facilitate land-based investments. The LAP has been a key enabler, helping to accelerate processes of land and agricultural commercialization in a context where small-scale farming still predominates. While land grabs in Ghana are less extensive than in Cambodia, there are reports of displacements where chiefs – all of whom are male – use their prerogative as custodians of customary land to sell plots, including communal lands, for their own private gain. Within an agricultural system based on small- to medium-sized farms of less than five hectares, the loss of even small amounts of farming land may have disastrous consequences for the livelihoods of rural people (Li, 2011).

Our research design also enabled comparisons within countries. In Cambodia, we selected six communes in the provinces of Ratanakiri, Kampong Thom, and Kratie. All were heavily affected by ELCs, but the magnitude and severity of impact varied for small landholders, in some cases, among villages of the same commune. The rapid pace of livelihood transformation in Cambodia also allowed for comparisons over time. Moreover, we were able to build on previous research by Gironde et al. (2015) in Ratanakiri, which provided long-term data for this case. In Ghana, we selected four districts, two in the Southern Region (Kwaebibirem and Asunafo North), and two in the North (East Gonja and Garu Tempane). This allowed for comparisons across space, capturing ecological and climatic differences as well as degrees of commercialization.

The project employed a range of methods, from legal and discourse analysis to econometrics and feminist geographies. We collected extensive data over the course of seven years, including close to 500 interviews with local communities, government officials, and civil society actors; and two rounds of household surveys in the four districts of Ghana and six communes of Cambodia.

Scope of the book

The chapters in this collection offer a selection of our results, organized along our research questions. The first section reports findings on the impacts of commercialization on livelihoods and the adequacy of food. The chapters illustrate how commercialization processes create unequal opportunities for female and male farmers in Ghana, and generate food insecurity and indebtedness in some households in Cambodia, with women responsible for dealing with both. They also show deteriorations in the quality of food and the privileging of men's nutritional needs over women's. The chapters in the second section examine the way policies and policy processes shape commercialization. They highlight how gender mainstreaming has been implemented in the policy areas of agriculture and food security in our case countries, illustrating the entanglement of an increasingly authoritarian political order with patriarchal politics in Cambodia and the tokenistic inclusion of women in agricultural policy-making in Ghana. They also discuss the co-optation of key government agencies by agribusinesses at the local level, their ability to push commercialization and cash crops at the expense of food crops, but also their focus on women's participation in agriculture. The third section examines the development and implementation of the right to food and gender equality through policies and laws in a context of commercialization. The Ghanaian case provides scope for reflection on the meanings ascribed to constitutional guarantees on gender equality and draws on recent jurisprudence to demonstrate the potential role of the judiciary in interpreting social development objectives as including the right to food. It also helps us to problematize the reality of institutional multiplicity and legal pluralism, that is the co-existence of customary and statutory law. In contrast, the Cambodian case provides materials to explore fierce conflicts around land management and gendered resistance to expropriation, and also allows us to trace the way in which gendered norms surrounding land inheritance are in flux as a result of commercialization. In the book's conclusion, we provide a comparative discussion of our findings in Cambodia and Ghana and argue that commercialization limits the availability, accessibility, and quality of food. The research documents violations of the right to food that are apparent in fluctuations in the seasonal availability of both food and income, gender, ethnic, age, and class-based barriers to accessing food and the prevalence of food that is of poor quality and limited cultural acceptability. To remedy these problems, we recommend that governments and agribusinesses prioritize the promotion and protection of the right to food and related rights to land, decent work, and social security.

Our project has sought to highlight a range of dimensions that are frequently overlooked or undervalued in studies of agricultural commercialization and its outcomes. These dimensions include food cultures, policy translation, national and subnational political and legal structures, the roles of multi-national agribusiness companies, national laws and constitutions, and the local practices and norms that construct gendered subjectivities. In presenting our findings, we hope to deepen the understanding of the impacts of commercialization in Cambodia and Ghana and offer guidance towards implementing rights-based policy approaches to agricultural development and food security globally.

References

Agarwal, B. (1994) *A Field of One's Own: Gender and Land Rights in South Asia.* Cambridge: Cambridge University Press.

Amanor, K. J. (2012) 'Global resource grabs, agribusiness concentration and the smallholder: two West African case studies', *The Journal of Peasant Studies*, 39(3–4), pp. 731–749, https://doi.org/10.1080/03066150.2012.676543.
Ashwood, L., Canfield, J., Fairbairn, M. and De Master, K. (2020) 'What owns the land: the corporate organization of farmland investment', *Journal of Peasant Studies*, https://doi.org/10.1080/03066150.2020.1786813.
Baglioni, E. and Gibbon, P. (2013) 'Land grabbing, large- and small-scale farming: what can evidence and policy from 20th century Africa contribute to the debate?', *Third World Quarterly*, 34(9), pp. 1558–1581, https://doi.org/10.1080/01436597.2013.843838.
Barndt, D. (2008) *Tangled Routes: Women, Work, and Globalization on the Tomato Trail*. Lanham, MD: Rowman & Littlefield.
Barrientos, S. et al. (1999) *Women and Agribusiness: Working Miracles in the Chilean Fruit Export Sector*. Houndmills, Basingstoke, New York: Macmillan.
Beban, A. and Bourke Martignoni, J. (2021) '"Now the forest is over": transforming the commons and remaking gender in Cambodia's uplands', *Frontiers in Sustainable Food Systems*, 5, pp. 1–15.
Behrman, J., Meinzen-Dick, R. and Quisumbing, A. (2012) 'The gender implications of large-scale land deals', *Journal of Peasant Studies*, 39(1), pp. 49–79, https://doi.org/10.1080/03066150.2011.652621.
Bello, W. (2008) 'Manufacturing a global food crisis', *Global Asia*, 3(2), pp. 15–25.
Bhattacharya, T. (2017) *Social Reproduction Theory: Remapping Class, Recentering Oppression*. London: Pluto Press.
Biddulph, R. (2010) Geographies of evasion: The development industry and property rights interventions in early 21st century Cambodia. Publications edited by the Departments of Geography, University of Gothenburg, Series B, no. 117. Gothenburg: Department of Human and Economic Geography, University of Gothenburg.
Bigler, C. et al. (2017) 'Rwanda's gendered agricultural transformation: a mixed-method study on the rural labour market, wage gap and care penalty', *Women's Studies International Forum*, 64, pp. 17–27, https://doi.org/10.1016/j.wsif.2017.08.004.
Borras, S. M. et al. (2011) 'Towards a better understanding of global land grabbing: an editorial introduction', *Journal of Peasant Studies*, 38(2), pp. 209–216, https://doi.org/10.1080/03066150.2011.559005.
Borras, S. M. and Franco, J. (2012) 'Global land grabbing and trajectories of agrarian change: a preliminary analysis', *Journal of Agrarian Change*, 12(1), pp. 34–59, https://doi.org/10.1111/j.1471-0366.2011.00339.x
Bourke Martignoni, J. (2018) 'Engendering the right to food? International human rights law, food security and the rural woman', *Transnational Legal Theory*, 9(3–4), pp. 400–414, https://doi.org/10.1080/20414005.2018.1568789.
Bourke Martignoni, J. (2020) 'The right to food', in Dugard, J., Porter, B., Ikawa, D. and Chenwi, L. (eds.), *Research Handbook on Economic, Social and Cultural Rights as Human Rights*. London: Edward Elgar, pp. 138–159.
Bourke Martignoni, J. (2021) 'A feminist methodology for implementing the right to food in agrarian communities: reflections from Cambodia and Ghana', *The Journal of Peasant Studies*, pp. 1459–1484, https://doi.org/10.1080/03066150.2021.1928642.
Bourke Martignoni, J. and Claeys, P. (2022) 'Without feminism there is no food sovereignty? Negotiating gender equality in the United Nations Declaration on the rights of peasants and other people working in rural areas', in Alabrese, M.,

Bessa, A., Brunori, M. and Giuggioli, P. F (eds.), *The United Nations' Declaration on Peasants' Rights*. Oxon: Routledge, pp. 47–61.

Brandt Broegaard, R., Vang Rasmussen, L., Dawson, N., Mertz, O., Vongvisouk, T. and Grogan, K. (2017) 'Wild food collection and nutrition under commercial agriculture expansion in agriculture-forest landscapes', *Forest Policy and Economics*, 84, pp. 92–101, https://doi.org/10.1016/j.forpol.2016.12.012.

Byres, T. J. (2012) 'The agrarian question and the peasantry', in Fine, B. and Saad-Filho, A. (eds.), *The Elgar Companion to Marxist Economics*. Cheltenham: Edward Elgar, pp. 10–15.

Calkin, S. (2015) 'Feminism, interrupted? Gender and development in the era of "Smart Economics"', *Progress in Development Studies*, 15(4), pp. 295–307, https://doi.org/10.1177/1464993415592737.

Carletto, C., Corral, P. and Guelfi, A. (2017) 'Agricultural commercialization and nutrition revisited: empirical evidence from three African countries', *Food Policy*, 67, pp. 106–118, https://doi.org/10.1016/j.foodpol.2016.09.020.

Carney, J. A. (2021) 'Subsistence in the Plantationocene: dooryard gardens, agrobiodiversity, and the subaltern economies of slavery', *The Journal of Peasant Studies*, 48(5), pp. 1075–1099, https://doi.org/10.1080/03066150.2020.1725488.

CESCR. (1999). 'The right to adequate food. General Comment 12'. UN Doc. E/C.12/1999/5.

Chu, J. (2011) 'Gender and "land grabbing" in Sub-Saharan Africa: women's land rights and customary land tenure', *Development*, 54(1), pp. 35–39, https://doi.org/10.1057/dev.2010.95.

Clapp, J. and Helleiner, E. (2012, May). 'Troubled futures? The global food crisis and the politics of agricultural derivatives regulation', *Review of International Political Economy*, 19(2), 181–207, https://doi.org/10.1080/09692290.2010.514528.

Clapp, J. and Fuchs, D. (eds.). (2009) *Corporate Power in Global Agrifood Governance*. Cambridge, MA: MIT Press.

Collier, P. (2008) 'The politics of hunger: How illusion and greed fan the food crisis', *Foreign Affairs*, November/December. Available at: http://www.foreignaffairs.com/articles/64607/paul-collier/the-politics-of-hunger.

Commission on the Status of Women. (2018) *Challenges and Opportunities in Achieving Gender Equality and the Empowerment of Rural Women and Girls. Agreed Conclusions of the 62nd Commission on the Status of Women*. New York: UN Women.

Cotula, L., Oya, C., Codjoe, E. A., Eid, A., Kakraba-Ampeh, M., Keeley, J. and Rizzo, M. (2014) 'Testing claims about large land deals in Africa: findings from a multi-Country study', *Journal of Development Studies*, 50(7), pp. 1–23.

Daley, E. and Pallas, S. (2014) 'Women and land deals in Africa and Asia: weighing the implications and changing the game', *Feminist Economics*, 20(1), pp. 178–201, https://doi.org/10.1080/13545701.2013.860232.

Daley, E. and Park, C. M. Y. (2012) *The Gender and Equality Implications of Land-related Investments on Land Access and Labour and Income-Generating Opportunities: A Case Study of Selected Agricultural Investments in Northern Tanzania*. Rome: FAO.

Daniel, S. (2011) 'Land grabbing and potential implications for world food security', in Behnassi, M. et al. (eds.), *Sustainable Agricultural Development. Recent Approaches in Resources Management and Environmentally-Balanced Production Enhancement*. Dordrecht: Springer, pp. 25–42.

Deere, C. D. and León, M. (eds.). (1987) *Rural Women and State Policy: Feminist Perspectives on Latin American Agricultural Development*. Boulder, CO, London: Westview Press.

Delarue, J. (2011) Thailand: the world's leading exporter of natural rubber owing to its smallholders, Working Paper No. 96 (Paris: Agence Française de Développement). Available at: http://www.afd.fr/webdav/site/afd/shared/PUBLICATIONS/RECHERCHE/Scientifiques/Documents-de-travail/096-document-travail-VA.pdf (accessed on 23 April 2015).

De Schutter, O. (2011) 'Green rush: the global race for farmland and the rights of land users', *Harvard International Law Journal*, 52, p. 503.

De Schutter, O. (2014) 'Final report of the special rapporteur on the right to food: the transformative potential of the right to food'. UN Doc. A/HRC/25/57.

De Schutter, O. (2015) 'The role of property rights in the debate on large-scale land acquisitions', *International Development Policy*, 6, https://journals.openedition.org/poldev/2026.

Diepart, J.-C. and Sem, T. (2018) *Cambodian peasantry and formalisation of land right: historical perspectives and current issues*. Research series. Paris: French Technical Committee on Land Tenure and Development (AFD-MEAE).

Doss, C., Meinzen-Dick, R. and Bomuhangi, A. (2014) 'Who owns the land? Perspectives from rural ugandans and implications for large-scale land acquisitions', *Feminist Economics*, 20(1), pp. 76–100, https://doi.org/10.1080/13545701.2013.855320.

Dufey, A., Vermeulen, S. and Vorley, W. (2007) *Biofuels: Strategic Choices for Commodity Dependent Developing Countries*. Amsterdam: Common Fund for Commodities.

Edelman, M., Oya, C. and Borras, S. M. (2013) 'Global land grabs: historical processes, theoretical and methodological implications and current trajectories', *Third World Quarterly*, 34(9), pp. 1517–1531, https://doi.org/10.1080/01436597.2013.850190.

Ellis, F. and Biggs, S. (2001) 'Evolving themes in rural development 1950s–2000s', *Development Policy Review*, 19, pp. 437–448, https://doi.org/10.1111/1467-7679.00143.

Elson, D. (1998) 'The economic, the political and the domestic: businesses, states and households in the organisation of production', *New Political Economy*, 3(2), pp. 189–208.

Esquivel, V. (2017) 'Efficiency and gender equality in growth theory: simply add-ons?', *Canadian Journal of Development Studies/Revue canadienne d'études du développement*, 38(4), pp. 547–552, https://doi.org/10.1080/02255189.2017.1377061.

Fakhri, M. (2020) 'Vision of the special rapporteur on the right to food'. UN Doc. A/HRC/46/33.

FAO (2006) 'Food security', *Policy Brief*, (2). Available at: http://www.fao.org/fileadmin/templates/faoitaly/documents/pdf/pdf_Food_Security_Cocept_Note.pdf (accessed on 10 August 2021).

FAO, IFAD, UNICEF, WFP and WHO (2021) *The State of Food Security and Nutrition in the World 2021. Transforming Food Systems for Food Security, Improved Nutrition and Affordable Healthy Diets for All*. Rome: FAO. https://doi.org/10.4060/cb4474en

Federici, S. (2019) *Re-Enchanting the World: Feminism and the Politics of the Commons*. Oakland, CA: PM Press.

Fox, J. and Castella, J.-C. (2013) 'Expansion of rubber (Hevea brasiliensis) in mainland Southeast Asia: what are the prospects for smallholders?', *The Journal of Peasant Studies*, 40(1), pp. 155–170, https://doi.org/10.1080/03066150.2012.750605.

Fraser, N. (2016) 'Contradictions of capital and care', *New Left Review*, 100, pp. 99–117.

Friedmann, H. and McMichael, P. (1989) Agriculture and the state system: the rise and decline of national agricultures, 1870 to the present. *Sociologia Ruralis*, 29(2), pp. 93–117.

Gironde, C., Golay, C. and Messerli, P. (2015) *Large-Scale Land Acquisitions: Focus on South-East Asia. Large-Scale Land Acquisitions*. Leiden: Brill Nijhoff.

Gironde, C., Reysoo, F., Torrico Ramirez, A. and Suon, S. (2021) 'No cash, no food. Gendered reorganization of livelihoods and food security in Cambodia', *The Journal of Peasant Studies*, 48(7), pp. 1485–1501, https://doi.org/10.1080/03066150.2021.1960826.

Gladwin, C. H. (ed.) (1991) *Structural Adjustment and African Women Farmers*. 1st edition. Gainesville: University of Florida Press.

Golay, C. (2011) 'Deux approches pour lutter contre la faim et la discrimination contre les femmes: le droit à l'alimentation et la souveraineté alimentaire', in Verschuur, C. (ed.) *Du grain à moudre. Genre, développment rural et alimentation*. Geneva: The Graduate Institute, pp. 341–359.

Gouyon, A. (1995) Paysannerie et hévéaculture dans les plaines orientales de Sumatra, unpublished PhD thesis (Paris: Institut National Agronomique Paris-Grignon).

Graziano Da Silva, J., Del Grossi, M.E. and Galvao de França, C. (eds.). (2011) *The Fome Zero (Zero Hunger) Program: The Brazilian Experience*. Brasilia: Ministry of Agrarian Development. Available at: http://www.fao.org/3/i3023e/i3023e.pdf.

Graziano Da Silva, J. (Coord) (2019) *From Fome Zero to Zero Hunger – A Global Perspective*. Rome: FAO, http://www.fao.org/3/ca5524en/CA5524EN.pdf.

Hajjar, R. et al. (2019) 'Capital, labor, and gender: the consequences of large-scale land transactions on household labor allocation', *The Journal of Peasant Studies*, 0(0), pp. 1–23, https://doi.org/10.1080/03066150.2019.1602520.

Hall, D. (2011). 'Land grabs, land control, and Southeast Asian crop booms', *Journal of Peasant Studies*, 38(4), pp. 837–857.

Hall, R., Scoones, I. and Tsikata, D. (2017) 'Plantations, outgrowers and commercial farming in Africa: agricultural commercialisation and implications for agrarian change', *The Journal of Peasant Studies,* 44(3), pp. 515–537, https://doi.org/10.1080/03066150.2016.1263187.

Harriss-White, B. (2005) 'Commercialisation, commodification and gender relations in post-harvest systems for rice in South Asia', *Economic and Political Weekly*, 40(25), pp. 2530–2542.

Hopma, J. and Woods, M. (2014) 'Political geographies of "food security" and "food sovereignty"', *Geography Compass*, 8(11), pp. 773–784, https://doi.org/10.1111/gec3.12163.

Ironside, J. (2009) 'Poverty reduction or poverty creation? A study on achieving the millennium development goals in two indigenous Communities in Ratanakiri Province, Cambodia', in Bourdier, F. (ed.), *Development and Dominion – Indigenous Peoples of Cambodia, Viêt Nam and Laos*. Bangkok: White Lotus Press, pp. 79–113.

Izumi, K. (2007) 'Gender-based violence and property grabbing in Africa: a denial of women's liberty and security', *Gender and Development*, 15(1), pp. 11–23.

Julia and White, B. (2012) 'Gendered experiences of dispossession: oil palm expansion in a Dayak Hibun community in West Kalimantan', *The Journal of Peasant Studies*, 39(3–4), pp. 995–1016, https://doi.org/10.1080/03066150.2012.676544.

Lastarria-Cornhiel, S. (1997) 'Impact of privatization on gender and property rights in Africa', *World Development*, 25(8), pp. 1183–1371.

La Via Campesina and allies (2007) Declaration of the forum for food sovereignty, Nyéléni, https://nyeleni.org/IMG/pdf/DeclNyeleni-en.pdf.

Levien, M. (2017) 'Gender and land dispossession: a comparative analysis', *The Journal of Peasant Studies*, 44(6), pp. 1111–1134, https://doi.org/10.1080/03066150.2017.1367291.

Li, T. M. (2011, March) 'Centering labor in the land grab debate', *The Journal of Peasant Studies*, 38(2), pp. 281–298, https://doi.org/10.1080/03066150.2011.559009.

Losch, B. (2004) 'Debating the multifunctionality of agriculture: from trade negotiations to development policies by the south', *Journal of Agrarian Change*, 4(3), pp. 336–360, https://doi.org/10.1111/j.1471-0366.2004.00082.x.

Luna, J. K. (2019) 'The chain of exploitation: intersectional inequalities, capital accumulation, and resistance in Burkina Faso's cotton sector', *The Journal of Peasant Studies*, 46(7), pp. 1413–1434, https://doi.org/10.1080/03066150.2018.1499623.

Maffii, M. (2009) 'Living on the margins: minorities and borderlines in Cambodia and Southeast Asia', in *Mainland Southeast Asia at its Margins: Minority Groups and Borders*. Siem Reap: Center for Khmer Studies. Available at: https://www.academia.edu/4687530/Living_on_the_Margins_Minorities_and_Borderlines_in_Cambodia_and_Southeast_Asia.

Margulis, M. E. (2013) 'The regime complex for food security: implications for the global hunger challenge', *Global Governance: A Review of Multilateralism and International Organizations*, 19(1), pp. 53–67, https://doi.org/10.1163/19426720-01901005.

Margulis, M. E., McKeon, N. and Borras, S. M. (2013) 'Land grabbing and global governance: critical perspectives', *Globalizations*, 10(1), pp. 1–23, https://doi.org/10.1080/14747731.2013.764151.

Matondi, P. B., Havnevik, K. and Beyene, A. (eds.). (2011) *Biofuels, Land Grabbing and Food Security in Africa*. London: ZedBooks.

Maye, D. and Kirwan, J. (2013) 'Food security: a fractured consensus', *Journal of Rural Studies*, 29, pp. 1–6, https://doi.org/10.1016/j.jrurstud.2012.12.001.

Mazoyer, M. and Roudart, L. (1997) *Histoire des agricultures du monde. Du néolithique à la crise contemporaine*. Paris: Éditions du Seuil.

McKeon, N. (2014) *Food Security Governance: Empowering Communities, Regulating Corporations*. London, New York: Routledge.

McMichael, P. (2013) 'Value-chain agriculture and debt relations: contradictory outcomes', *Third World Quarterly*, 34(4), pp. 671–690, https://doi.org/10.1080/01436597.2013.786290.

McMichael, P. (2020) 'Does China's 'going out' strategy prefigure a new food regime?' *The Journal of Peasant Studies*, 47(1), pp. 116–154, https://doi.org/10.1080/03066150.2019.1693368.

Menon, N. (2015, April 25) 'Is feminism about "women"?' *Economic and Political Weekly*, L(17), pp. 37–44.

Meyers, W. H. and Meyer, S. (2008) Causes and implications of the food price surge, Food and Agricultural Policy Research Institute (FAPRI), University of Missouri–Columbia.

Mies, M. (2012) *The Lace Makers of Narsapur*. Melbourne: Spinifex Press.

Monsalve Suárez, S. (2013) 'The human rights framework in contemporary agrarian struggles', *The Journal of Peasant Studies*, 40(1), pp. 239–290, https://doi.org/10.1080/03066150.2011.652950.

Mutopo, P. (2011) 'Women's struggles to access and control land and livelihoods after fast track land reform in Mwenezi District, Zimbabwe', *The Journal of Peasant Studies*, 38(5), pp. 1021–1046, https://doi.org/10.1080/03066150.2011.635787.

Mutopo, P. and Chiweshe, M. (2014) 'Large-scale land deals, global capital and the politics of livelihoods: Experiences of women small-holder farmers in Chisumbanje, Zimbabwe', *International Journal of African Renaissance Studies – Multi-, Inter- and Transdisciplinarity*, 9(1), pp. 84–99, https://doi.org/10.1080/18186874.2014.916879.

Naidu, S. C. and Ossome, L. (2016) 'Social reproduction and the agrarian question of women's labour in India', *Agrarian South: Journal of Political Economy*, 5(1), pp. 50–76.

Oya, C. (2013) 'The land rush and classic agrarian questions of capital and labour: a systematic scoping review of the socioeconomic impact of land grabs in Africa', *Third World Quarterly*, 34(9), pp. 1532–1557, https://doi.org/10.1080/01436597.2013.843855.

Paramita, P. (2013). Between Chamkar and the Kitchen: a Livelihood Approach to the Implication of Land Grabs on Food Security in Cambodian Rural Households. *Master diss., Graduate Institute of International and Development Studies.* http://www. snis. ch/system/files/paramita_mdev_dissertation_2013_1_0. pdf.

Park, C. M. Y. and Maffii, M. (2017) '"We are not afraid to die": gender dynamics of agrarian change in Ratanakiri province, Cambodia', *The Journal of Peasant Studies*, 44(6), pp. 1235–1254, https://doi.org/10.1080/03066150.2017.1384725.

Park, C. M. Y. and White, B. (2017) 'Gender and generation in Southeast Asian agro-commodity booms', *The Journal of Peasant Studies*, 44(6), pp. 1103–1110, https://doi.org/10.1080/03066150.2017.1393802.

Piacenza, C. (2012) 'Neogtiating gendered property relations over land: oil palm expansion in Kalangala District, Uganda', in *International Conference on Global Land Grabbing II*. Ithaca, NY: Cornell University.

Pierce Colfer, C. J. (2013) *The Gender Box: A Framework for Analysing Gender Roles in Forest Management*. Bogor: Center for International Forestry Research.

Pingali, P. L. and Rosegrant, M. W. (1995) 'Agricultural commercialization and diversification: processes and policies', *Food Policy*, 20(3), pp. 171–185, https://doi.org/10.1016/0306-9192(95)00012-4.

Polanyi, K. (1944) *The Great Transformation*, Holt, Rinehart et Winston, Inc. Reprint 1975. New York: Octagon Books.

Prügl, E. (2016) 'Neoliberalism with a feminist face: crafting a new hegemony at the World Bank', *Feminist Economics*, 23(1), pp. 30–53, https://doi.org/10.1080/13545701.2016.1198043.

Prügl, E. and Joshi, S. (2021) 'Productive farmers and vulnerable food securers: contradictions of gender expertise in international food security discourse', *The Journal of Peasant Studies*, 48(7), pp. 1439–1458, https://doi.org/10.1080/03066150.2021.1964475.

Razavi, S. (ed.) (2003) *Agrarian Change, Gender and Land Rights*. Oxford: Blackwell Publishing.

Razavi, S. (2007) 'Liberalisation and the debates on women's access to land', *Third World Quarterly*, 28(8), pp. 1479–1500, https://doi.org/10.1080/01436590701637342.

Razavi, S. (2009) 'Engendering the political economy of agrarian change', *The Journal of Peasant Studies*, 36(1), pp. 197–226.

Roberts, A. and Soederberg, S. (2012). 'Gender equality as "smart economics"? A critique of the 2012 World Development Report', *Third World Quarterly*, 33(5), pp. 949–968, https://doi.org/10.1080/01436597.2012.677310.

Rulli, M. C. and D'Odorico, P. (2014) 'Food appropriation through large scale land acquisitions', *Environmental Research Letters*, 9(6), n.p., http://doi.org/10.1088/1748-9326/9/6/064030.

Saab, A. (2019) *Narratives of Hunger: Feeding the World in Times of Climate Change*. Cambridge University Press.

Sans, P. and Combris, P. (2015) 'World meat consumption patterns: an overview of the last fifty years (1961–2011)', *Meat Science*, 109, pp. 106–111, https://doi.org/10.1016/j.meatsci.2015.05.012.

Scott, J. W. (1986, December) 'Gender: a useful category of historical analysis', *The American Historical Review*, 91(5), pp. 1053–1075, https://doi.org/10.2307/1864376.

Sen, A. (1981) *Poverty and Famines: An Essay on Entitlement and Deprivation*. Oxford, New York: Clarendon Press, Oxford University Press.

Sikor, T. (2012) 'Tree plantations, politics of possession and the absence of land grabs in Vietnam', *Journal of Peasant Studies*, 39(3–4), pp. 1077–1101, https://doi.org/10.1080/03066150.2012.674943.

Smaller, C., & Mann, H. (2009). *A Thirst for Distant Land. Foreign Investment in Agricultural Land and Water*. International Institute for Sustainable Development, pp. 48-59.

Spieldoch, A. and Murphy, S. (2009) 'Agricultural land acquisitions: implications for food security and poverty alleviation', in Kugelman, M., Levenstein, S. (eds.), *Land Grab? The Race for the World Farmland*. Washington, DC: Woodrow Wilson International Center for Scholars, pp. 39–53.

Suarez Franco, A. (2021) 'The human right to food: lessons learned toward food systems transformation', in Kjaerum, M., Davis, M. and Lyons, A. (eds.), *Covid-19 and Human Rights*. London: Routledge, pp. 191–208.

Sulle, E. and Dancer, H. (2020) 'Gender, politics and sugarcane commercialisation in Tanzania', *The Journal of Peasant Studies*, 47(5), pp. 973–992, https://doi.org/10.1080/03066150.2019.1632294.

The State of Food Security and Nutrition in the World 2021. FAO, IFAD, UNICEF, WFP and WHO, 2021. https://doi.org/10.4060/cb4474en.

Titche, A. (2017) 'Land grabs and food security: the international community should adopt a code of conduct to protect local communities and improve global food security', *U.C. Irvine Law Review*, 7, p. 473.

Tsikata, D. (2009) 'Gender, land and labour relations and livelihoods in sub-Saharan Africa in the era of economic liberalisation: towards a research agenda', *Feminist Africa*, 12, pp. 11–30.

Tsikata, D. (2016) 'Gender, land tenure and agrarian production systems in Sub Saharan Africa', *Agrarian South: Journal of Political Economy*, 5(1), pp. 1–19, https://doi.org/10.1177/2277976016658738.

Tsikata, D. and Amanor-Wilks, D. E. (2009) 'Editorial: Land and labour in gendered livelihood trajectories', *Feminist Africa*, 12, pp. 1–10.

Tsikata, D. and Yaro, J. (2014) 'When a good business model is not enough: land transactions and gendered livelihood prospects in rural Ghana', *Feminist Economics*, 20(1), pp. 202–226.

van der Ploeg, J. D. (2008) *The New Peasantries: Struggles for Autonomy and Sustainability in an Era of Empire and Globalization*. Sterling, VA: Earthscan.

White, B., Borras Jr. S. M., Hall, R., Scoones, I. and Wolford, W. (2012) 'The new enclosures: critical perspectives on corporate land deals', *The Journal of Peasant Studies*, 39(3–4), pp. 619–647.

Whitehead, A. (2008) 'The gendered impacts of liberalisation policies on african agricultural economies and rural livelihoods', in Razavi, S. (ed.), *The Gendered Impacts of Liberalization: Towards 'Embedded Liberalism'*. London: Routledge, pp. 37–62.

World Bank. (1986) *Poverty and Hunger: Issues and Options for Food Security in Developing Countries*. Washington, DC: The International Bank for Reconstruction and Development/The World Bank.

World Food Summit: Plan of Action (1996) World Food Summit, 13–17 November 1996, Rome, Italy. Available at: http://www.fao.org/wfs/index_en.htm.

Ziegler, J. et al. (2011) *The Fight for the Right to Food: Lessons Learned*. Springer.

Zoomers, A. (2010). 'Globalisation and the foreignisation of space: seven processes driving the current global land grab', *The Journal of Peasant Studies*, 37(2), pp. 429–447. https://doi/ 10.1080/03066151003595325

Section I

Commercialized livelihoods, gender, and food security

Dzodzi Tsikata

Introduction

How does the commercialization of land and agriculture affect the food security and right to food of smallholder farmers? How are gender, class, generation, and location implicated in the processes and outcomes of land and agricultural commercialization? And how have smallholder farmers responded to livelihood shocks that threaten their food security and right to food? These are the three broad questions that animate this section. Constituted by four chapters, two each on Cambodia and Ghana, the section contributes to the polarized debates about agricultural commercialization and food security through micro-level studies that examine how processes of land and agricultural commercialization have unfolded in particular communities and how smallholders have responded to processes of land concentration and conversion and to their dispossession or adverse incorporation into circuits of accumulation. All four studies are framed by three or four measures of food security and the right to food (availability, accessibility, adequacy/utilization, and stability) and a gender analysis of access to and control over productive resources and livelihood responses.

Each chapter compares different regions in the country, thus drawing attention to how gender interacts with class, generation, and agroecology to determine food security outcomes of agricultural commercialization. Two of the studies (Dzanku and Tsikata; Gironde et al.) are based on two rounds of surveys in Kampong Thom and Ratanakiri provinces in Cambodia, and Asunafo North, Kwaebibirem, East Gonja, and Garu-Tempane districts in four regions of Ghana with common questions between 2016 and 2020. In the case of Cambodia, an earlier study (2011–2013) provided important continuity for observations of certain trends in rural livelihoods. The other two chapters (Reysoo; and Eweh and Tsikata) are based on qualitative studies undertaken in 2015 before the surveys and in between surveys.

What separates the Cambodia and Ghana cases is that they study sites at very different stages of land and agricultural commercialization. The

DOI: 10.4324/9781003202004-2

Cambodia sites have been at the frontier of momentous agrarian change in the last 20 years. As Reysoo notes in her contribution, the economic, social, and cultural changes in frontier areas occur at an extraordinary pace and can be observed as they unfold. This has enabled the Cambodia papers to plot the distinctive pathways of change in household production systems and the local economy, the changing roles of men and women, and the growing social differentiation among local and migrant communities. The Ghana research sites, on the other hand, are undergoing long durée steady processes of smallholder commercial agriculture that appear quite settled in terms of dynamics of change. The two chapters therefore focus on examining the livelihood outcomes and food cultures of land and agricultural commercialization occurring along different timelines.

Gironde et al. and Dzanku and Tsikata both found that improvements in food availability and food diversity notwithstanding, smallholder households were increasingly dependent on food markets for their food supply, and this had implications for the access and adequacy/utilization and stability aspects of food security. In both countries, households continued to experience seasonal food shortages and volatility in food prices. Gironde et al. found food insecurity with clear gender differences in access and utilization in Cambodia, with the situation worse in Ratanakiri, which had fewer avenues for wage employment than Kampong Thom. The Ghana survey findings were more ambiguous: while Dzanku and Tsikata found clear gender differences in access to productive resources and the gross value of crop sales, the levels of commercialization between male and female households and men and women were similar in all the study communities. This, they argue, is likely due to distress sales on the part of female-headed households. In relation to food security, Dzanku and Tsikata found differences between male- and female-headed households in terms of availability of food throughout the year, but no difference in crops cultivated and dietary diversity in the matrilineal, more commercialized, tree-crop-dominated forest zones of southern Ghana. In the patrilineal, less commercialized, food-crop-dominated savannah zones of northern Ghana, however, there were significantly more diverse diets for male-headed households and clear differences in the crops grown by men and women.

We might surmise that the differences in the food security findings between Cambodia and Ghana are linked to the fact that more women in Cambodia had completely exited own account agricultural production and were working as agricultural labourers or food vendors, unlike in the Ghana cases, where many more male- and female-headed households in the sample continued to produce some of their own food. Secondly, the visibility and significance of rural food markets is something that is taken for granted in rural Ghana with its longer traditions of cooked food vending by female itinerant and stationary traders. These differences in findings could also be a question of the meaningfulness of the measures of food security, a point both Dzanku and Tsikata and Gironde et al. draw attention to. For example, household food security measures do not address intra-household differences, and the food diversity measure does not account for differences in the quantities of a particular food group consumed by households and their different members. These limitations of quantitative measurements of food security are addressed somewhat through the discussion of food cultures and practices in the four chapters.

Our understanding of food insecurity is further deepened by Reysoo's mapping of the processes by which women lost farmlands and had to turn to independent food sales and low-paid wage work that do not make up for food self-provisioning, as well as her finding that, as with other economic activities, the food trade is gender-segmented, with women at the bottom of the pile in terms of the size and capital base of their operations. Eweh and Tsikata also draw attention to the ways in which different aspects of food cultures map onto the various dimensions of food security in an approach that sheds light on the food cultures of agricultural commercialization. Furthermore, both Reysoo and Eweh and Tsikata found that the reliance on cooked food, while advantageous in terms of time savings and convenience, was also more expensive and less nutritious or fresh, with implications for the access and adequacy dimensions of food security.

In conclusion, our studies reinforce the literature that finds that, in contexts of land and agricultural commercialization, the pathways to food insecurity are not singular, and they can affect different dimensions of food security in gendered and location-specific ways that re-inscribe gender orders, even as some of their contours might change. This has implications for the work to promote gender-equitable food security and the right to food in different contexts of land and agricultural commercialization.

1 From food crop to food shop. Agricultural commercialization, food security, and gender relations in Cambodia

Christophe Gironde, Andres Torrico Ramirez, Amaury Peeters, and Kim Thida

Introduction

Over the last two decades, the highlands of Cambodia have experienced an unprecedented territorial expansion and intensification of commercial agriculture driven by large-scale land acquisitions and significant migration from the lowlands. Rural livelihoods have changed radically in terms of access to natural resources, land tenure, economic activities, and the way people procure their food. Non-edible cash crops and wage labour have become the pillars of the household economy. Borrowing money has become crucial for the purchase of farming inputs and equipment, consumption goods, housing, and motorbikes, as well as for buying food.

Overall, the food security situation in Cambodia has improved over the last 20 years, in the sense that the country has not experienced 'drastic food insecurity' (Culas and Tekreo, 2016). However, the Food and Agriculture Organization (FAO) considers that Cambodia is characterized by 'serious hunger' (FAO, 2019) given that undernourishment is lying between 15% and 31%. Moreover, undernutrition as well as insufficient dietary diversity and quality to meet nutrient needs remain high (McDonald et al., 2015; Talukder et al., 2010). Lastly, there is a growing concern about food safety and nutrition. Food availability remains vulnerable to climate and natural disasters, and price fluctuations have a serious impact on the poor in terms of access to food.

New livelihood systems have also induced changes in gender relations, with regard to the organization of labour, decision-making, and responsibilities. These changes in turn contribute to reshape gender orders and hierarchies (Gironde et al., 2021). This chapter addresses the outcomes of this process with regard to gender relations and food security. It aims at answering two main questions. The first one is what have been the major changes induced by the reorganization of livelihoods with regard to food security, including procurement, availability, access, and diversity of food? And to what extent do these outcomes differ in terms of gender? The

DOI: 10.4324/9781003202004-3

second main question is what strategies do households deploy to adapt to the new environment and to what extent are these strategies gendered? Our analysis covers a period of around seven years (2013–2020), based on field research carried out in the two provinces of Ratanakiri and Kampong Thom in 2011–2013 and between 2016 and 2020.

The first section after this one recalls the discussions around the expansion of commercial agriculture, its consequences on food security, and how gender relations may be implicated in food security. The second section presents the analytical framework, the data, and the methods used. Then, we introduce the research areas and the major socio-economic changes that have affected them. Research findings are presented in the next two sections, which include an analysis of food security and household strategies with a systematic gender consideration. The final section discusses our findings and their contribution to the broader understanding of the interrelations between agricultural commercialization, gender relations, and food security.

Agricultural commercialization, food security, and gender

The global land rush and the expansion of large-scale farming units is predominantly seen as a threat to the food security of impacted communities, as many lose lands that were previously used for food production. The situation may be aggravated as most of the land acquisitions are turned into plantation farming of non-edible crops and/or food crops that are exported (Titche, 2017; Rulli and D'Odorico, 2014).

Over the last 20 years, Cambodia has experienced such a process, including a strong development of rubber, cassava for animal feed, sugarcane, and cashew. The shift to cash crops has accelerated with the set-up of economic land concessions (ELCs) measuring thousands of hectares (Peeters, 2015; Oldenburg and Neef, 2014) and other landholdings of dozens to hundreds of hectares (Gironde et al., 2015). As a consequence, the majority of rural populations have seen their access to land eroded (Diepart and Sem, 2016; Un and So, 2011). Traditional ways of life, including free access to natural resources from forest, grazing, and water areas that were crucial to subsistence and swidden farming have been undermined (Frieson, 2010; Bourdier, 2009). Small landholders have no choice but to reorganize their cropping systems according to crop booms (Mahanty and Milne, 2016; Bourdier, 2012). This process is found overall to have negative consequences on impacted communities (Hak et al., 2018; Bühler et al., 2015; Jiao et al., 2015). Cash-crop price volatility has exposed small producers to severe consequences that are best illustrated by the fall of cassava prices (Hought et al., 2012). In addition, cash-crop incomes have turned out not to be up to the levels of investments, making many producers indebted (Gironde and Ramirez, 2019; Mahanty and Milne, 2016; Bylander, 2015).

The reconfiguration of agriculture and food procurement systems must take gender relations into account, considering the different and disproportionate impacts on women when compared to men (Levien, 2016; Joshi, 2015; Behrman et al., 2014) and the significant contribution of women to food production and family food management (Agarwal, 2011). There has been a broad consensus that the transformation of rural livelihoods in Cambodia has resulted in a loss of status and increased vulnerability of women when compared to the socio-economic organization that prevailed before the development of the market economy as depicted by Ebihara and Mertha (1965), White (1996), and Bourdier (2009). Frieson (2001) dates this back to the Democratic Kampuchea era and the war, including a culture of violence and inequality in access to land that did not exist before. The status of rural women is deemed to have further eroded, as the importance of subsistence agriculture, in which women were key, has been shrinking (Park and Maffii, 2017). Gender inequality is considered to increase also since men have taken new and predominant roles in commercial transactions, wage labour, and interactions with outsiders, while women have tended to remain confined to the domestic sphere (Joshi, 2020; Kusakabe, 2015; Paramita, 2013). However, several authors have drawn attention to the uneven situations of women who should not be considered as a homogeneous group (Frewer, 2017; Levien, 2017).

Our research addresses more particularly the consequences of gender relations for food security. First, food security is deemed correlated with the proportion of land controlled by women (Scanlan, 2004). In this regard, it has been argued about land grabs that women overall lose access to more land compared to men (Julia and White, 2012). Second, women who are considered to put more emphasis on food crops than their husbands may lose their traditional role in providing food because of the growing importance of cash crops (Holden and Ghebru, 2016). Third, the increasing workload of women leaves them insufficient time to devote to preparing food for the family (Komatsu et al., 2018; Johnston et al., 2015). Four, women are considered to better prioritize spending on food items and health over non-essential items (Seymour et al., 2019; Fan and Pandya-Lorch, 2012).

Analytical framework and data

Our approach to food security builds on the FAO definition of 'food and nutrition security', which includes availability, access, stability, and quality of food for all people in all seasons (FAO, 2012). Availability of food is approached through the way people procure their food. Access is documented throughout dietary diversity, food shortages, and food shocks. Stability is addressed throughout the duration and seasonality of food shortage and by crossing shortage with income from wage labour along the year.

We build on the livelihood system framework (Scoones, 2015) to address the change in food procurement. Among the various productive resources, we analyse in particular households' access to natural capital, i.e., land for farming and natural resources from the wild, and access to financial capital, which is key for the purchase of agricultural inputs. Access to financial capital, including borrowing from other households, credit purchases in shops, and bank loans, is also of growing importance for the purchase of food items. We examine households' strategies, in this case their capability to adapt crop choices according to market demand and the respective prices of farming inputs, outputs, and food items. This allows us to highlight how producers make trade-offs between subsistence and cash crops as well as own-produced and purchased food. To avoid a productive resources-production determinism, we use the concept of 'assets' of Bebbington (1999) according to whom resources such as land are not only 'a means to make a living' (1999: 2022) but also 'the basis of agent's power to act' (idem). This is combined with Bylander's concept of 'negotiated decision-making', which allows to highlight the respective agency of household members, economic constraints, and gender norms (Bylander, 2014:13).

Gender differences are approached through the lenses of access to and control over productive resources, i.e., outputs in kind from farming and cash income from wage labour. Women may have the same access as men to productive resources but not the same control over them, i.e., the power to decide how these resources are used. These powers reflect historically constructed attributes and identities, codes of conduct, and expected behaviours (Brickell, 2011; Frieson, 2010; Ledgerwood, 1996). We analyse how these translate and are adapted into everyday life practices and their evolution induced by the reorganization of livelihoods. Throughout access and control, we highlight changes in food procurement and food security and the extent to which they are influenced by men and women's respective workload and responsibilities, as well as women's position relative to men (March et al., 2005). Our gender analysis is combined with the livelihood framework to analyse how access to productive resources is subject to change (Peluso and Ribot, 2020) in the context of deforestation, land-use change, new technologies, migration, the increasing need for cash, and the development of market, which all contribute to reshaping productive strategies and social norms.

The field research used both quantitative and qualitative methods. Two questionnaire-based surveys were carried out, including 200 randomly selected households in each of the two provinces of Kampong Thom and Ratanakiri in 2016 and 700 and 459 households respectively in 2019. The design of questionnaires and semi-structured interview guidelines were adapted to address intra-household gender relations, including specific questions to be answered by women. The surveys were programmed on KoBoToolbox and were implemented through mobile phones with the use of the KoBoCollect app. This allowed us to save time and prevent mistakes at the time of implementation. The dataset was later processed and analysed with Stata.

In parallel, a series of 430 semi-structured interviews with household members and around 50 with local political authority representatives were conducted between 2016 and 2019. In addition, many unplanned interviews were conducted with people we met by chance in the villages and in marketplaces in the neighbourhood of the selected sites. We coded the transcripts of the interviews in NVivo; a node-list based on the topics included in the questionnaire was developed with the aim that quantitative findings would be compared, substantiated, and illustrated by the findings from semi-structured interviews. Interviewees were anonymized, and statements are not credited to any specific individual or public authority. Quotes in the chapter are translations of original ones made in Khmer or in the language of indigenous communities (mostly Tampouan and Jarai). We also used data and information from previous research carried out between 2011 and 2013 in the same sites.

The changing landscape of Kampong Thom and Ratanakiri

In Kampong Thom, the four selected communes, Bong Lvea, Kraya, Prasat, and Thnoat Chum used to rely primarily on wet rice in the lower lands and on cassava in the upper lands. Farming was complemented by fishing, buffalo raising, the sale of various resources from the forest (bamboo, rattan, resin), and logging. Livelihoods changed radically with the development of ELCs in the second half of the 2000s, which attracted newcomers who settled along the roads and cleared lands. Cashew and cassava developed, as transportation infrastructure improved and traders became more numerous. Logging reduced as rubber companies occupied forest areas. Casual wage labour also developed as the newcomers and households who managed to accumulate lands were in need of labour force.

In Ratanakiri, the two studied communes, Loum Choar and Malik, were up until the mid-1990s characterized by the relative abundance of land for swidden farming, mostly rain-fed rice and vegetables in a rotating system including 10–15 years of fallow land after 2–3 years of use of the same plot. Populations also used to rely on free access to forests and rivers that were rich in game, food, and medicinal plants. Cash crops, such as soybeans and cashews, were introduced during the 1990s. Rubber developed later with the sharp increase of its price in 2005.

The development of ELCs and other medium-sized landholdings has consequently impacted the five selected communes in terms of loss of access to resources from the wild and land for farming. Overall, the losses were more severe in Kampong Thom, including cases of forced displacement and relocation 12 km away on sites where families received only residential land plots with poor soil quality (Paramita, 2013). In comparison, in Ratanakiri, the development of ELCs was less severe given the abundance of land. Land losses were however very uneven between villages from the same commune depending on the proximity to ELCs. On the other side, the availability of land and the rubber boom have attracted far more numerous migrants to

Ratanakiri. The non-native population represents about one third of the total population; the majority of them acquired lands and some of them developed small- and medium-sized rubber plantations ranging from two to five hectares to several dozen hectares.

The change in landscape in the two studied areas is characterized first by the reduction of the forest and other areas for fishing and grazing once they were acquired by ELCs and others. In 2016, 92% of our sampled population reported that free access to forest areas had reduced over the previous five years; 87% for access to water areas for fishing and 86% in the case of grazing areas. The percentages are similar in 2019, slightly more severe in the case of Kampong Thom, slightly less in Ratanakiri where some of the ELCs have not developed as planned. Furthermore, natural resources in the wild have been seriously depleted.

A second major change is the reduction of households' land for farming, from 2.9 ha in 2016 to 2.5 ha in 2019 in Kampong Thom and from 4.5 to 4.2 ha in the case of Ratanakiri. The share of households with less than two ha and less than one ha has markedly increased in Kampong Thom. The trend is the same but is at a slower pace in Ratanakiri (Table 1.1).

Differences in landholding are significant across generations. In 2019, the mean land area of households with the head being below 25 years of age is 2.1 ha compared to 3.2 ha for the whole sample. When compared to 2016, the gap across generations has been on the rise. We did not find evidence that women lost more access to land than men when looking at landholding: lands are jointly held and farmed by wife and husband. There also is no evidence of more control of the use of land by the man versus by the woman. Joint decision and working together are predominant and even slightly on the rise. With the change in natural environment and land use, men-specific tasks, such as slashing trees and hunting, have reduced and been eased, thanks to chainsaws and hand-tractors. The tasks that were women-specific have also changed, thanks to new technologies such as brush-cutters, weedicides, and fertilizers. Some gender division of labour persists, such as ploughing by men, but most of the farming tasks can now be carried out by women as well as men (Gironde et al., 2021).

Another major change is the development of wage labour, mostly farm work on a daily basis for other households and medium-sized plantations.

Table 1.1 Percentage of households with less than 1 ha and 2 ha

	Kampong Thom		*Ratanakiri*	
	2016 (%)	*2019 (%)*	*2016 (%)*	*2019 (%)*
Less than 2 ha	44	56	23	27
Less than 1 ha	27	38	12	13

Source: Demeter Surveys (2016, 2019)

Both men and women have increasingly engaged in wage labour, with a higher participation of men who are often preferred over women. This preference from employers relates to physical strength and the search for hired workers with tools and machines, such as chainsaws and hand-driven tractors, which are used more by men than women (Gironde et al., 2021). Additionally, women's lower participation in wage labour is due to the fact that they are occupied with reproductive activities.

Lastly, the studied communes are characterized by their increasing interconnection with the outside, including the development of commercial activities and people movement. Trade and service activities have also developed in administrative, provincial, and district towns, including commercial banks that have become a major player in the transformation of the rural economy. The two studied areas have enjoyed a growing number of marketplaces and shops selling food in district centres, along the roads and even in the smallest and most remote villages. Trade has developed with the tarring of some roads and the opening of new ones. Nowadays, every household either has a motorbike or can borrow one from a relative, and enjoy more options with regard to where to buy food. Food purchases have also developed, thanks to the possibility of buying on credit from the shops.

Shopping has become by far the main way people procure their food (Figure 1.1). The importance of food purchase is particularly significant in Kampong Thom where, in 2016, 84% of our sample reported that purchase was the main source for procuring their rice; they were 91% and 98% reporting the same respectively for fish and fruits.

Overall, the importance of food purchases is lower in Ratanakiri where the reduction of access to natural resources was less severe. However, in these areas that had forests rich in game and where there was plenty of space for grazing, it is noticeable that 90% of the households responded that purchase was the main way they procured their meat. The same applies for fish with 72% of the sample reporting that the main source of procurement is purchased whereas people used to fish in surrounding water streams.

Food security

When interviewing people on the issue of food and change over time, the large majority states first that 'now there is more', 'it's better than before', and 'now food is available all year round'. However, any discussion around food rapidly moves to food price, inflation, and the inability to afford such and such items, or the need to borrow. In 2019, like in 2016, the most reported shock in the two provinces was the *increase in the price of major food items consumed.*

Access, stability, and diversity

While availability has increased, access to food remains a serious challenge as shown by the evolution and duration of food shortages. The two studied

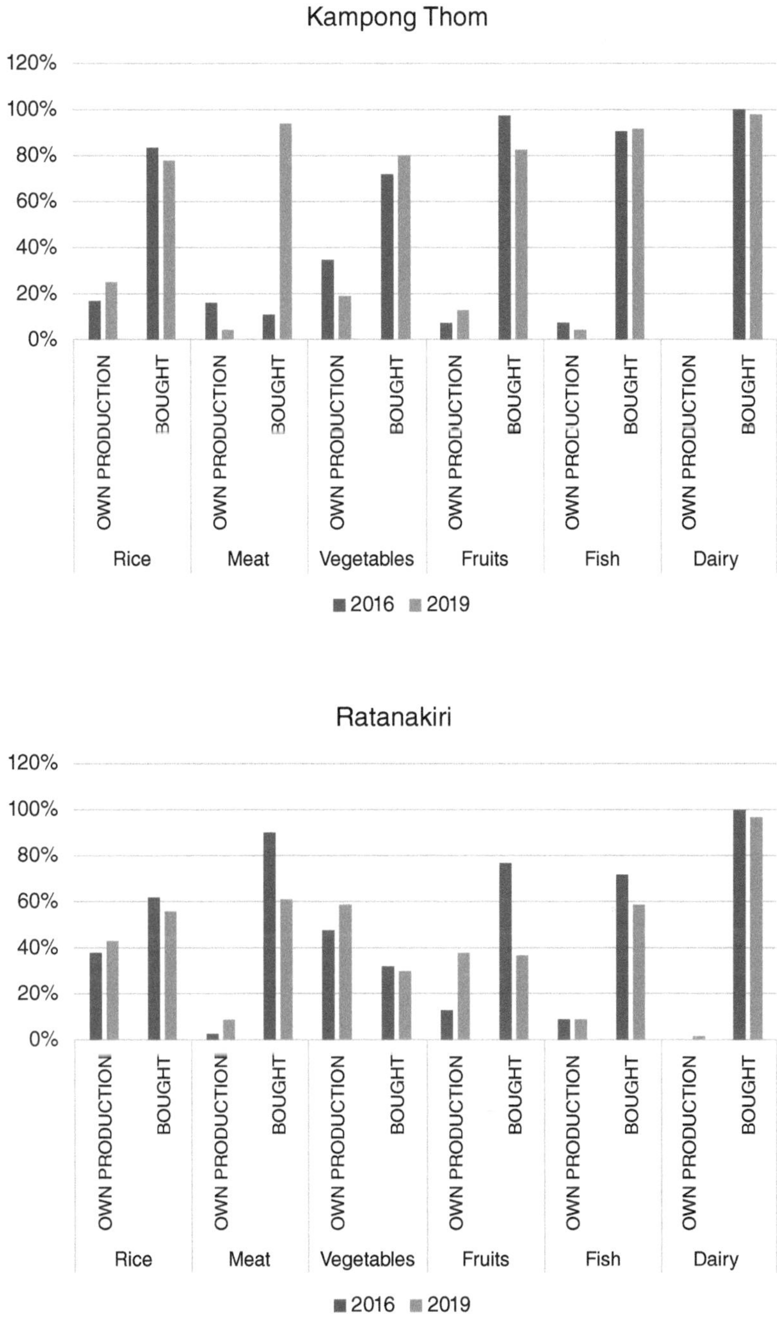

Figure 1.1 Main source of food procurement
Source: Demeter Surveys (2016, 2019).

areas present opposite trends with regard to food shortages (see Table 1.2). In Kampong Thom, the percentage of households affected by food shortage over the last 12 months fell from 57% in 2016 to 26% in 2019, and the mean duration of the shortage reduced from two-and-half months to one month. In contrast, in Ratanakiri, this percentage increased from 38% to 63%, and the duration of food shortage doubled. The trends are the same for extreme food shortage, which has reduced by half in Kampong Thom down to 7% but has increased substantially up to 30% in Ratanakiri.

A major reason for these differences between the studied areas is the access to wage labour and non-farm self-employment. In 2019, 86% of households in Kampong Thom have income from wage labour and/or non-farm-self-employment whereas there are only 45% in the case of Ratanakiri. This correlation between food shortages and income from wage employment in particular can be observed across seasons. Figure 1.2 shows that food shortage

Table 1.2 Food shortage and duration over the last 12 months

	Kampong Thom		*Ratanakiri*	
	2016	*2019*	*2016*	*2019*
Percentage of HH who reported food shortage	57%	26%	38%	63%
Average number of months of food shortage	2.5	1.1	1.28	2.5
Percentage of HH who reported extreme food shortage	15%	7%	17%	30%
Average number of months of extreme food shortage	0.6	0.3	0.4	1

Source: Demeter Surveys (2016, 2019)

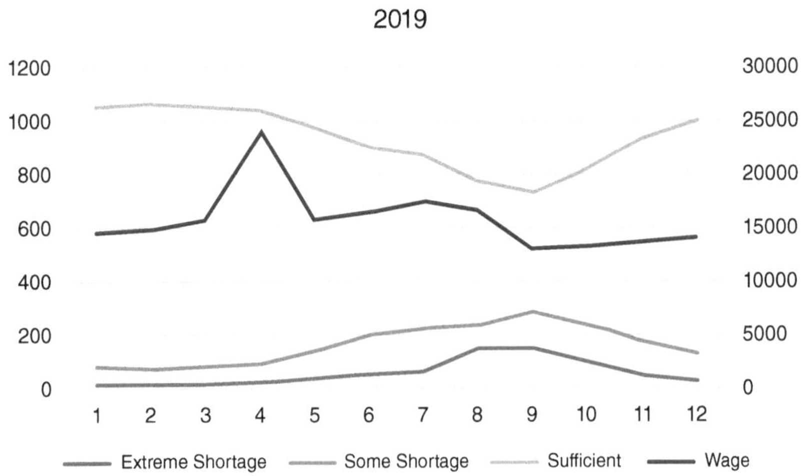

Figure 1.2 Food shortage and wages over time
Source: Demeter Surveys (2016, 2019)

Table 1.3 Percentage of households with '*poor*' and '*borderline*' food consumption scores by land size

	0.5–2 ha (%)	*2–5 ha (%)*	*5–7 ha (%)*
Kampong Thom	6	4	2
Ratanakiri	42	30	19

Source: Demeter Surveys (2016, 2019)

reaches its peak in the month of September whereas the number of days in waged employment reaches its lowest point. Conversely, in the months of December through March, wage employment is at its highest point and food shortages are at their lowest, and since wage employment in the local market is predominantly in agriculture, cash and food shortages are closely linked to the agricultural cycle.

Lastly, we assessed access to food through food diversity. We found in 2019 that 22% of the sample in Ratanakiri had a *borderline* food consumption score (FCS) – seven days before the survey – and 3% had a *poor* FCS, against 5% and 1% respectively in Kampong Thom. The two provinces differ also with regard to food dietary diversity measured by the household dietary diversity score (HDDS). It is also interesting to note that overall dietary diversity as measured by the HDDS for the last 24 hours has not changed between 2016 and 2019 in either of the two provinces. In order to understand the dynamics behind food security, we analysed the relationship between FCS and land size. We found that households with less land are more prone to being food insecure as shown in Table 1.3. This relationship appears to be stronger in Ratanakiri, where 42% of households with 0.5–2 ha reported having an FCS level of poor or borderline compared to 19% of those with 5 ha or more. In Kampong Thom, we find a similar but less pronounced trend, which could be due to a more widespread access to alternative economic activities, including wage employment, and thus less reliance on land.

Coping, cooking, and shopping: women's affair

The common general statements that there are more diverse food products available must be put into perspective: food supply is undoubtedly more diversified, but that does not mean that food diet is. We did not find significant gender differences in terms of the types of food consumed (during the last 24 hours) except for alcohol and tobacco, which were consumed 2.5 times more by men than women. Also, men eat outside far more often than women do. For the rest, during our interviews addressing the organization of meals, food intake, and types, interviewees repeatedly explained that 'everyone eating together eats the same'. However, dietary diversity indicators conceal actual intra-household dynamics. Hence, we analysed gender differences in coping strategies in case of food shortages and found significant differences. Adult

Table 1.4 Household members' adaptation in case of food shortage

	Adult female only (%)	*Adult male only (%)*	*Both adults (%)*	*All members (%)*
Who reduced the number of meals eaten per day?				
Kampong Thom	43	0	30	27
Ratanakiri	23	14	45	18
Who reduced the portion size of meals?				
Kampong Thom	43	0	33	24
Ratanakiri	16	8	59	18
Who relied on less preferred, less expensive food?				
Kampong Thom	26	3	41	29
Ratanakiri	9	8	44	35

Source: Demeter Surveys (2016, 2019)

women bear a heavier burden than men with regard to the responsibility of finding coping mechanisms to face these challenges (Table 1.4). The gender gap is particularly striking in the case of Kampong Thom, where 43% of households reported that it is only women who reduced the number of meals eaten per day in situations of lack of food, whereas the percentage is zero for men only. The situation is similar for the reduction of portion size and when it comes to who relies on less preferred and less expensive food. In Ratanakiri, the gender differences are much lower, revealing that coping mechanisms are shared in a more equal manner.

Our interviews reveal that the rationale for, and legitimization of, these adaptations is the energy needed to go to work: 'household members who go to the field need three meals per day and those that stay home, two meals per day'. This conception benefits men more than women, especially in Kampong Thom where women tend to dedicate significantly less time to paid work compared to men (4.5 h vs. 7 h) and more hours to unpaid domestic work (4.2 h vs. 1.3 h). The difference is even higher (6.3 h vs. 1.6 h) when taking into account the time spent caring for children. These differences are lower in Ratanakiri, where the gap is only 0.8 h on average. The difference in unpaid work is 2 h and 3.5 h when including care work. This has important ethnic implications. When looking at ethnicity and time-use, we find that the gap in terms of time spent working is smaller for indigenous households and larger for non-indigenous households. In indigenous households, women play an important role in paid work, particularly in agriculture, which in turn has implications in terms of food intake as women exhibit a 'need of energy' similar to that of men. However, we still find that even though they spend more time than men working overall (paid and unpaid labour), they are still the main contributors to food security and are responsible for facing food shortages.

Table 1.5 Time spent on meal preparation (in hours and minute)

	Kampong Thom		*Ratanakiri*	
	2016	*2019*	*2016*	*2019*
Both	1 h 49 m	1 h 31 m	1 h 58 m	1 h 31 m
Male	31 m	7 m	46 m	13 m
Female	1 h 17 m	1 h 23 m	1 h 13 m	1 h 17 m

Source: Demeter Surveys (2016, 2019)

When discussing change in food diet and food preparation, a common statement from many interviewees is that there is less time to prepare food, leading to a lower-quality diet. This is exemplified by the purchase of dried noodles instead of eating self-produced rice, and the consumption of biscuits, soda, and ice-cream by the children. Our surveys support the time reduction statement when we consider both adults (Table 1.5). However, this reduction is noticed in the case of men, while the time women spent on meal preparation has slightly increased.

Interviewees explain that when they work for someone else, at the end of the day, they cannot bring any vegetables back home as they do when they work on their own land, nor can they bring fish caught in traps in surrounding water streams. Furthermore, hired workers return later than when they work on their farm. They may lack the time to then go to buy food along the main roads and markets that are outside the villages. Even if the food market is a few kilometres from home, a motorbike makes a big difference in time compared to a bicycle. And although every household owns a motorbike, for the many owning only one, it happens that the woman cannot ride to the market in case the man is using it. Another factor often evoked is the fatigue from the working day: 'buying food is easier', meaning it is not as tiring as preparing fresh products. In such situations, some resort to ready-made food from peddlers or grocery shops inside the village. However, if village shops are highly convenient, there is less choice and more industrial products, and they are a bit more expensive than bigger shops or rural markets outside the villages. In any case, purchasing food reduces a bit of the effort but does not make a big difference in the time needed to prepare meals. Meat or fish is bought already cooked, but rice must be cooked at home. Also, buying dried noodles or ready-made soups has not become a regular way of eating. It is rather occasional and often to feed one member rather than the whole household. All these accommodations help out occasionally, but do not fundamentally change the time spent cooking.

The increasing consumption of purchased food is also a matter of agenda. With the diversification of productive activities, activities outside and more circulation, household members have different schedules. So, families tend to eat less together as they do when they all return together at the end of the day spent on their farm. Some return later, and some eat outside. This is an additional challenge for feeding the whole family, and women can count less on men to address it.

When it comes to decision-making about spending, our data indicates that men remain predominant in decision-making for the biggest expenses and loans, typically buying motorbike and farming equipment. Meanwhile, women hold more decision-making power for everyday expenses including food as illustrated by the answers from this 45-year-old man interviewed in Ratanakiri:

QUESTION: Who makes the decision to buy food from outside?
ANSWER: My wife does.
Q: What are the three things for which you are the main decision-maker?
A: Selling and buying crops, constructing houses, and buying motorbikes.
Q: For what things can your wife make decisions without having to consult with you?
A: She can buy food and provide paddy rice or milled rice to relatives.

In the 2019 survey, 62% of women responded that they single-handedly manage food expenditures and 56% responded that they single-handedly decide the amount spent on food. When adding the decisions made 'mostly by woman but in consultation with someone else', the proportions reach up to 80%. There is a broad consensus, among women as among men, that it is better that it is the women who manage food purchases. Men recognize that, overall, women pay more attention to food: 'They buy only food and face less temptation to spend money for other things. Men can spend part of the money for coffee or a drink'. This recognition also reflects codes of conduct: it is expected from women not to waste money, whereas there is tolerance for men. Thus, discussions around deciding what food to buy and how much to spend indicate that women enjoy agency rather than power, as purchase options are limited by financial resources anyway.

Food procurement strategies

Shifting back to food crops

Our data reveals a shift in cropping systems from a rise in cash crops towards a renewed interest in food crops. This is particularly significant in Kampong Thom where the share of food crops in the total area increased from 13% in 2016 to 34% in 2019, alongside a reduction in the share of cash crops from 87% to 66%. The share of rice in farmed areas has almost tripled from 11% to 29%, whereas the share of cassava reduced from 57% to 35%. Farmers explained this change by the constant fall in the price of cassava since 2016; some reported that they did not even get the means for paying contracted hired labour. They then reconsidered their cassava-for-cash strategy and opted to some extent for the production of rice for self-consumption, or cashew as its price is more stable. Our data on food procurement confirms this trend, showing that the share of rice that was consumed from own-production increased from 17% in 2016 to 25% in 2019.

In Ratanakiri, we do not see a return to food crops of the same magnitude. The share of food crops increased only slightly from 14% to 16% and the one of cash crops reduced from 86% to 84%. However, it must be recalled that, prior to 2016, the share of rice had fallen to 10% of the total farmed area. In 2016, none of the respondents in Ratanakiri reported growing rain-fed rice. Rice was grown only in the low-wetland (*srey*) where no other crop can be grown. Since then, farmers have reconsidered their crop choice for the same reasons as in Kampong Thom, which is the falling price of cassava. They opted for cashew, as new varieties that grow faster were disseminated. The renewed interest in rice is found among those farmers who hold low wetlands. They also explained they started growing paddy rice again for their own consumption as they are concerned about the health and safety of purchased rice, in particular rice imported from Vietnam, which is suspected to be 'full of chemicals'. Even so, these households are the better-off, as they hold low wetlands and have enough income not to be obliged to buy the cheapest priced lowest-quality imported rice. For many households, the rice harvest is not enough to meet their needs.

The reorientation of farming towards more food crops raises the question of whether it has a gender drive, based on the argument that women put more emphasis on food crops than their husbands. When looking at who is the main decision-maker for what to grow on each farmed plot, we found a slight increase in man–woman joint decision-making between 2016 and 2019 (from 68% to 73% of the total farmed area, see Table 1.6). However, when disaggregating by type of crop, this number broke down to 66% when the crops grown were food crops and 79% when they were cash crops. In turn, 20% of households reported that women alone decided what to grow in the case of food crops compared to only 7% in the case of cash crops. We can conclude that, in general women, have a higher participation in decisions regarding what crops to grow, especially when it comes to food crops.

Similarly, women stand out as the main decision-makers for vegetables, where they decide on 31% of the plots against 14% for men. Moreover, women also hold more power for selling vegetables, being the main decision-makers in 48% of the cases against 12% for men, whereas for other crops, joint decision-making prevails for more than 70% of the cases.

Table 1.6 Decision-making by crops and gender

	All crops (%)	*Food crops (%)*	*Cash crops (%)*
Both	73	66	79
Female only	13	20	7
Male only	14	14	14

Source: Demeter Survey (2019)

Borrowing for food

Our surveys show that between 2016 and 2019, the amount borrowed for food increased 5.7-fold. At both extremes, big amounts (e.g. 300 kg of rice) can be paid in several instalments in big shops in town, and small amounts (for daily intake) can be bought on credit in small village shops. The two provinces present different patterns of change (see Table 1.7). In Kampong Thom, the percentage of households who reported borrowing for food in the last 12 months slightly decreased from 12% in 2016 to 10% in 2019 but the mean amount of loans for food increased three-fold, from US$678 in 2016 to US$2,083 in 2019. In Ratanakiri, the percentage of households who reported borrowing for food in the last 12 months increased from 26% in 2016 to 32% in 2019, and the mean amount of loans was multiplied by 1.6, from US$1,228 in 2016 to US$2,165 in 2019.

The difference in trends between the two study sites can be paralleled to the respective change in cropping system, i.e. an increase in food crops and corollary reduction of food purchases in Kampong Thom, which did not occur in Ratanakiri. The increase in the amounts borrowed for the purpose of buying food may also be linked to the increase of the price of food that people have to purchase. Furthermore, borrowing money has become a coping mechanism in case of food shortage. In Ratanakiri, 65% of the households who suffered food shortage in 2019 revealed that they borrowed money to cope with this situation. The percentage is 43% in the case of Kampong Thom. In Kampong Thom, we find that the percentage of households who reported borrowing for food in the last 12 months remains the same between the two time periods (around 10%). However, when asking about food shortages and coping mechanisms, we find that 21% of respondents faced a lack of food or water, and out of this subsample, 43% reported borrowing money as a response.

Women play a far more important role than men when it comes to borrowing to cope with lack of food, as shown in Table 1.8.

Table 1.7 Loans for the purpose of buying food

	Kampong Thom		*Ratanakiri*	
	2016	*2019*	*2016*	*2019*
Percentage of head of household who borrowed for food	12%	10%	26%	32%
Mean amount of loan in US$ for food/household	678	2,083	1,228	2,165
Mean amount of loan in US$ for all items/household	2,005	5,922	2,254	5,894
Mean share of food in total amount of loans	16%	10%	29%	32%

Source: Demeter Surveys (2016, 2019)

Table 1.8 Coping strategies and gender

	Adult female only (%)	*Adult male only (%)*	*Both adults (%)*
Who borrowed food or relied on help from friends or relatives?			
Kampong Thom	78	8	14
Ratanakiri	42	16	42
Who purchased food on credit or borrowed money to purchase food?			
Kampong Thom	80	2	18
Ratanakiri	42	16	42

Source: Demeter Survey (2019)

Back to the forest

Historically, natural resources played a key role in food procurement and, in particular, in case of a bad rice harvest. Access to the forest, water, and grazing areas has drastically reduced with the appropriation of those areas. And natural resources have been largely depleted. However, our research shows that they remain important for the most vulnerable population groups. The situation and the trend differ significantly between the two provinces when it comes to the practice of collecting food items from the wild and fishing. In Kampong Thom, the percentage of respondents who reported going to the forest to collect food products for consumption was 10% in 2016 and even reduced to 6% in 2019. A similar trend is found with regard to fishing: in 2016, 64% of respondents said they had gone fishing in the past 12 months; in 2019, only 32% did. There is simply not much left of natural resources in these areas.

Conversely, in Ratanakiri, there is not only a higher percentage, but above all, a strong increase in the number of people who report collecting food products from the forest: from 49% in 2016 up to 70% in 2019. This might be surprising when considering the reduced access to forests, but illustrates the fact that a fraction of the population who do not produce enough food and cannot afford to buy enough food resort to natural resources. They explain that they go to the forest when they have no other option to get food, typically when they do not find the opportunity to sell their labour. Figures show clearly that the purpose of food products collection is almost exclusively consumption and not sale. The percentage of respondents who sold in 2019 is 6% for food items from the forest and 8% in the case of fish. Actually, our qualitative interviews indicate that people resort to going to the forest or to going fishing although they do not get a lot. The territorial expansion of rubber tree plantations and the increasing presence and circulation of people have contributed to a reduction in game and fish.

As for gendered differences in terms of forestry and fishing, Table 1.9 shows that most of these activities are pursued predominantly by men, except for collecting food for consumption where women also play an important

Table 1.9 Use of forest resources and fishing by gender

	Cut trees for sale or home consumption (%)	*Collected food products for consumption (%)*	*Collected products for medicinal care (%)*	*Fished for family consumption (%)*	*Fished for commercial purposes (%)*
Male only	67	30	70	80	71
Female only	5	18	10	3	18
Both	28	52	20	17	11

Source: Demeter Survey (2019)

role. This goes in line with our previous findings that the responsibilities of procuring food and coping strategies lie with the women.

Discussion and conclusion

The argument that the increasing commercialization of agriculture is a threat to food security is overall supported by our findings. The territorial expansion of large-scale farming and non-edible cash crops, first, has largely deprived populations of the food they could procure from the wild, which was important in terms of diversity and stability as well as in the event of insufficient harvests. Second, for most producers, cash crops provide income that is not up to the investments and expenses and more so because households' landholdings have diminished in size. Those who do not manage to compensate by selling their labour force suffer increasing food shortages and for longer periods. As a result, purchasing enough food of good quality remains out of reach for many. Borrowing has become crucial, not only for productive activities, but also for consumption. Thus, the analysis of the impact of agricultural commercialization on food security must be refined, in this case by taking into account non-farm income that has become crucial (Do et al., 2019), as demonstrated by the difference between the studied areas.

The outcomes of this broader process of agrarian change in terms of food security differ according to gender. The proportion of women experiencing smaller and fewer meals as well as going to bed hungry is higher than that of men. If the types of food consumed are overall the same for men and women, gender differences are remarkable in case of food shortage. Women reduce food consumption far more than men, and this occurs even in cases where they work more than men. The role of women must thus also be assessed through their greater effort/sacrifice compared to men, as also shown by Reysoo and Suon (2017).

We find that the argument according to which women's status was eroded must be rethought. First, the comparison over time is not simple, given the magnitude of change in livelihoods since the expansion of the market economy. Subsistence agriculture has diminished, but this has not reduced the importance of women. They play a major role and have decision-making power in food production but also in the everyday tasks of cooking, especially as men contribute less than before. It is also women's daily duty to buy food and eventually borrow money for feeding the family. Discussions around deciding what food to buy and how much to spend indicate that women enjoy agency rather than power (Bylander, 2014), as purchase options are limited by financial resources. Moreover, women often lack time to prepare food and have to resort to purchased food items because of their increasing workload. Time spent for making income competes with time dedicated to care work. In sum, the major role they play is synonymous with responsibility and stress rather than power or status. Yet, in any case, they prioritize expenditures on food better than men, who are unanimous to acknowledge that.

This chapter shows that the reconfiguration of gender relations induced by the commercialization of agriculture and its respective outcomes for women and men are complex, and vary according to various dimensions and not just land and cropping system. The same is true of the consequences for food security. New vulnerabilities have developed, including price volatility, stability of income, and food quality. They are undeniably gendered as it is the women who bear the responsibility for feeding the family. However, in a context of increasing socio-economic differentiation, gender relations and food security must be further analysed across class as well as ethnicity and social norms (Clement et al., 2019; Frewer, 2017).

Contrary to the view of agricultural commercialization as an irreversible process, our research shows that small landholders attempt to revert to more self-produced food. For the well-off, this trend also reflects a growing concern about the quality of purchased food. However, this is not possible for everyone and overall, self-produced food is far below food needs. Given the state of depletion of foodstuffs that people could freely access and the pressure to continue cash crops, food stores have a bright future. The fact that it is the women who are responsible for borrowing raises the question as to whether they may bear more or specific responsibility, stress, and pressure to maintain good social relations and the reputation of the household.

References

Agarwal, B. (2011) *Food Crises and Gender Inequality*, DESA Working Paper No. 107. https://www.un.org/esa/desa/papers/2011/wp107_2011.pdf

Baird, I. (2014) 'The Global Land Grab Meta-Narrative, Asian Money Laundering and Elite Capture: Reconsidering the Cambodian Context', *Geopolitics*, 19(2), pp. 431–453. https://doi.org/10.1080/14650045.2013.811645

Bebbington, A. (1999) 'Capitals and Capabilities: A Framework for Analyzing Peasant Viability, Rural Livelihoods and Poverty', *World Development*, 27(12), pp. 2021–2044. https://doi.org/10.1016/S0305-750X(99)00104-7

Behrman, J., Meinzen-Dick, R. and Quisumbing, A. (2014) 'An Interpretation of Large-Scale Land Deals Using Boserup's Theories of Agricultural Intensification, Gender and Rural Development'. In: Fischer-Kowalski, M., Reenberg, A., Schaffartzik, A., and Mayer, A. Eds. *Ester Boserup's Legacy on Sustainability.*

Human-Environment Interactions, vol. 4. Springer, Dordrecht. https://doi.org/10.1007/978-94-017-8678-2_12

Bourdier, F. (2009) *Ethnographie des populations indigènes du nord-est cambodgien. La montagne aux pierres précieuses (Ratanakiri).* Paris: L'Harmattan.

Bourdier, F. (2012) 'Nouvelle agriculture et transformations sociales sur les hauts plateaux du Nord-Est cambodgien', *Autrepart*, 62, pp. 217–235. https://doi.org/10.3917/autr.062.0217

Brickell, K. (2011) 'We Don't Forget the Old Rice Pot When We Get the New One": Discourses on Ideals and Practices of Women in Contemporary Cambodia', *Journal of Women in Culture and Society*, 36(2), pp. 437–462. http://www.jstor.org/stable/10.1086/655915

Bühler, D. et al. (2015) *Rural Livelihood Strategies in Cambodia: Evidence from a Household Survey in Stung Treng*, ZEF Working Paper Series 137, Bonn, University of Bonn. https://www.econstor.eu/bitstream/10419/111695/1/820641170.pdf

Bylander, M. (2014) 'Contested Mobilities: Gendered Migration Pressures Among Cambodian Youth', *Gender, Place & Culture: A Journal of Feminist Geography*. https://doi.org/10.1080/0966369X.2014.939154

Bylander, M. (2015) 'Credit as Coping: Rethinking Microcredit in the Cambodian Context', *Oxford Development Studies*, 43(4), pp. 533–553. https://doi.org/10.1080/13600818.2015.1064880

Clement, F. et al. (2019) 'From Women's Empowerment to Food Security: Revisiting Global Discourses through a Cross-Country Analysis', *Global Food Security*, 23, pp. 160–172. https://doi.org/10.1016/j.gfs.2019.05.003

Culas, R. J. and Tek, K. (2016) 'Food Security in Cambodia: Trends and Policy Objectives', *International Journal of Development Issues*, 15(3), pp. 306–327. https://doi.org/10.1108/IJDI-06-2016-0033

Demeter Surveys (2016, 2019), https://zenodo.org/record/6539149

Diepart J.-C. and Sem, T. (2016) 'Fragmented Territories: Incomplete Enclosures and Agrarian Change on the Agricultural Frontier of Samlaut District, North-West Cambodia', *Journal of Agrarian Change*, 18(1), pp. 156–177. https://doi.org/10.1111/joac.12155

Do, T. L., Nguyen, T. T. and Grote, U. (2019) 'Nonfarm Employment and Household Food Security: Evidence from Panel Data for Rural Cambodia', *Food Security*, 11, pp. 703–718. https://doi.org/10.1007/s12571-019-00929-8

Dzanku, F. M., Tsikata, D. and Ankrah, D. A. (2021) 'The gender and geography of agricultural commercialization: what implications for the food security of Ghana's smallholder farmers?', *The Journal of Peasant Studies,* 48(7), pp. 1507-1536. https://doi.org/10.1080/03066150.2021.1945584

Ebihara, M. and Mertha, A. (1965) (Ed.) and ProQuest (Firm) (2018) *Svay: A Khmer Village in Cambodia.* Ithaca, NY: Southeast Asia Program Publications, an imprint of Cornell University Press.

Fan, S. and Pandya-Lorch, R. (Ed.) (2012) *Reshaping Agriculture for Nutrition and Health.* IFPRI 2020 book, Washington, DC: International Food Policy Research Institute.

FAO (2012) 'Committee on World Food Security, Thirty-Ninth Session', Rome. http://www.fao.org/3/MD776E/MD776E.pdf

Frieson, K. C. (2001) 'In the Shadows: Women, Power and Politics in Cambodia', Centre for Asia-Pacific Initiatives Occasional Paper 26.

Frieson, K. C. (Ed.) (2010) *Social Landscapes and Rural Livelihoods: Cambodian Communities in Transition.* Phnom Penh: The Learning Institute.

Frewer, T. (2017). 'The Gender Agenda: NGOs and Capitalist Relations in Highland Cambodia', *Critical Asian Studies*. https://doi.org/10.1080/14672715.2017.1300861

Gironde, C., Golay, C. and Messerli, P. (Eds.) (2015) 'Large-Scale Land Acquisitions: Focus on South-East Asia', *International Development Policy*. Geneva: Graduate Institute Publications, Boston: Brill-Nijhoff. https://doi.org/10.4000/poldev.2015

Gironde, C., Reysoo, F., Torrico Ramirez, A. and Suon, S. (2021) 'No Cash, No Food. Gendered Reorganization of Livelihoods and Food Security in Cambodia', *The Journal of Peasant Studies*. https://doi.org/10.1080/03066150.2021.1960826

Gironde, C. and Torrico Ramirez, A. (2019) "Dépossession foncière, transition agraire et capacité d'adaptation: Devenir des populations autochtones de Ratanakiri (Cambodge)', *Revue internationale des études du développement*, 238(2), pp. 291–322. https://doi.org/10.3917/ried.238.0291

Hak, S., McAndrew, J. and Neef, A. (2018) 'Impact of Government Policies and Corporate Land Grabs on Indigenous People's Access to Common Lands and Livelihood Resilience in Northeast Cambodia', *Land*, 7(4). https://doi.org/10.3390/land7040122

Holden, S. T. and Ghebru, H. (2016) 'Land Tenure Reforms, Tenure Security and Food Security in Poor Agrarian Economies: Causal Linkages and Research Gaps', *Global Food Security*, 10, pp. 21–28. https://doi.org/10.1016/j.gfs.2016.07.002

Hought, J. et al. (2012) 'Biofuels, Land Use Change and Smallholder Livelihoods: A Case Study from Banteay Chhmar, Cambodia', *Applied Geography*, 34, pp. 525–532. https://doi.org/10.1016/j.apgeog.2012.02.007

Jiao, X., Smith-Hall, C. and Theilade, I. (2015) 'Rural Household Incomes and Land Grabbing in Cambodia', *Land Use Policy*, 48, pp. 317–328. https://doi.org/10.1016/j.landusepol.2015.06.008

Johnston, D. et al. (2015) *Agriculture, Gendered Time Use, and Nutritional Outcomes: A Systematic Review*, International Food Policy Research Institute, Discussion Paper 01456.

Joshi, D. (2015) 'Gender Change in the Globalization of Agriculture?', *Peace Review*, 27(2), pp. 165–174. https://doi.org/10.1080/10402659.2015.1037620

Joshi, S. (2020) 'Working Wives: Gender, Labour and Land Commercialization in Ratanakiri, Cambodia', *Globalizations*, 17(1), pp. 1–15.

Julia and White, B. (2012) 'Gendered Experiences of Dispossession: Oil Palm Expansion in a Dayak Hibun Community in West Kalimantan', *Journal of Peasant Studies*, 39(3–4), pp. 995–1016. https://doi.org/10.1080/03066150.2012.676544

Kem, S. (2017) *Commercialisation of Smallholder Agriculture in Cambodia: Impact of the Cassava Boom on Rural Livelihoods and Agrarian Change*, PhD Thesis, University of Queensland, Australia.

Komatsu, H., Malapit, H. J. L. and Theis, S. (2018) 'Does Women's Time in Domestic Work and Agriculture Affect Women's and Children's Dietary Diversity? Evidence from Bangladesh, Nepal, Cambodia, Ghana, and Mozambique', *Food Policy*, 79, pp. 256–270. https://doi.org/10.1016/j.foodpol.2018.07.002.

Kusakabe, K. (2015). '*Gender Analysis of Economic Land Concessions in Cambodia and in Northern Laos: Case of Rubber Plantations*'. Conference Paper, Land Grabbing, Conflict and Agrarian – Environmental Transformations: Perspectives from East and Southeast Asia, 5–6 June 2015, Chiang Mai University. https://www.eur.nl/sites/corporate/files/CMCP_43-Kusakabe.pdf

Ledgerwood, J. (1996) 'Politics and Gender: Negotiating Conceptions of the Ideal Woman in Present Day Cambodia', *Asia Pacific Viewpoint*, 37(2), pp. 139–152.

Levien, M. (2017) 'Gender and Land Dispossession: A Comparative Analysis', *The Journal of Peasant Studies*, 44(6), pp. 1111–1134. https://doi.org/10.1080/03066150.2017.1367291

March, C. Smyth, I. and Mukhopadhyay, M. (1999) *A Guide to Gender-Analysis Frameworks*. Oxford: OXFAM.

Mahanty, S. and Milne, S. (2016) 'Anatomy of a Boom: Cassava as a "Gateway" Crop in Cambodia's North-Eastern Borderland', *Asia Pacific Viewpoint*, 57(2), pp. 180–193. https://doi.org/10.1111/apv.12122

McDonald, C. et al. (2015) 'Household Food Insecurity and Dietary Diversity as Correlates of Maternal and Child Undernutrition in Rural Cambodia', European Journal of Clinical Nutrition, 69, pp. 242–246. https://doi.org/10.1038/ejcn.2014.161

Oldenburg, C. and Neef, A. (2014) 'Reversing Land Grabs or Aggravating Tenure Insecurity? Competing Perspectives on Economic Land Concessions and Land Titling in Cambodia', *Law and Development Review*. https://doi.org/10.1515/ldr-2014-0014

Paramita, P. (2014) *Between Chamkar and the Kitchen: A Livelihood Approach to the Implication of Land Grabs on Food Security in Cambodian Rural Households*, Graduate Institute, MA-thesis Development Studies, Geneva.

Peeters, A. (2015) 'Disclosing Recent Territory – Lift and Rural Development Contributions of Cambodian Large – Scale Land Acquisitions'. Conference Paper, Land Grabbing, Conflict and Agrarian – Environmental Transformations: Perspectives from East and Southeast Asia, 5–6 June 2015, Chiang Mai University. https://www.eur.nl/sites/corporate/files/CMCP_79-Peeters.pdf

Park, C. and Maffii, M. (2017) '"We Are Not Afraid to Die": Gender Dynamics of Agrarian Change in Ratanakiri Province, Cambodia', *The Journal of Peasant Studies*, 44(6), pp. 1235–1254. https://doi.org/10.1080/03066150.2017.1384725

Peluso, L. N. and Ribot, J. C. (2020) 'Postscript: A Theory of Access Revisited', *Society & Natural Resources*, 33(2), pp. 300–306. https://doi.org/10.1080/08941920.2019.1709929

Reysoo, F. and Suon, S. (2017). '"In the Forest We Had Plenty": Gender, Food Culture and Diet Change in Rural Cambodia (Kampong Thom Province, 1995–2015)'. In Bock, B. and Duncan, J. Eds. *Gendered Food Practices from Seed to Waste*, Yearbook of Women's History 36. Hilversum: Uitgeverij Verloren, pp. 31–44.

Rulli, M. C. and D'Odorico, P. (2014) 'Food Appropriation through Large Scale Land Acquisitions', *Environmental Research Letters*, 9(6). http://doi.org/10.1088/1748-9326/9/6/064030

Scanlan, S. J. (2004) 'Women, Food Security, and Development in Less Industrialized Societies: Contributions and Challenges for the New Century', *World Development*, 32(11), pp. 1807–1829.

Scoones, I. (2015) *Sustainable Livelihoods and Rural Development*. Rugby: Practical Action Publishing.

Seymour, G., Masuda, J., Williams, J. and Schneider, K. (2019) 'Household and Child Nutrition Outcomes among the Time and Income Poor in Rural Bangladesh', *Global Food Security*, 20, pp. 82–92. https://doi.org/10.1016/j.gfs.2019.01.004

Talukder, A. et al. (2010) 'Homestead Food Production Model Contributes to Improved Household Food Security and Nutrition Status of Young Children and Women in Poor Populations', *Field Actions Science Reports* [Online]. https://journals.openedition.org/factsreports/404

Titche A. 2017. 'Land Grabs and Food Security: The International Community Should Adopt a Code of Conduct to Protect Local Communities and Improve Global Food Security', *UC Irvine Law Review*, 7, p. 473.

Un, K. and So, S. (2011) 'Land Rights in Cambodia: How Neopatrimonial Politics Restricts Land Policy Reform', *Pacific Affairs*, 84(2), pp. 287–306. https://doi.org/10.5509/2011842289

White J. (1996) 'The Indigenous Highlanders of the Northeast: An Uncertain Future', *Interdisciplinary Research on Ethnic Groups in Cambodia: Final Draft Reports*. Phnom Penh: Center for Advanced Study, pp. 333–373.

2 Gender, agricultural commercialization, and food security in Ghana

Fred Mawunyo Dzanku and Dzodzi Tsikata

Introduction

Agricultural commercialization has been an important driver of change in Ghana's political economy, pre- and post-independence. Colonial economic interests encouraged the commercialization of cocoa, oil palm, and rubber for export (Berry and Berry, 1975) and created a smallholder-dominated commercial agriculture system. Contemporary agricultural policies have continued to promote smallholder agricultural commercialization on the assumption that this would lead to economic development, income growth, and improvement in well-being. This has been, in part, responsible for the rapid growth in land commercialization.

Agricultural commercialization influences and is influenced by access to and control over the means of production (productive and reproductive resources) such as land, labour, technology, and the sexual division of labour. At the same time, access to and control over these resources are often unequal across different groups defined by class, gender, generation, and kinship. It is known that, since the colonial era, Ghana's agricultural production systems and commercialization have been shaped by gender dynamics, including gender-based division of productive and reproductive labour (Richards, 1983). The cash crop revolution of the early 20th century, for example, was highly gendered (Hopkins, 1973).

An important path through which agricultural commercialization in Ghana has been gendered is differences in crops produced by men and women. Women have a long history of heavy involvement in food crop production while men tend to dominate the non-food cash crop sector (Grosz-Ngaté, 1997), although changes have been observed (Lambrecht et al., 2018). Aside from crop choice, there are also significant differences in women's and men's access to land, labour, and agricultural capital (Dzanku et al., 2021). These gender differences have implications for the livelihoods of men and women. An important question, therefore, is what gendered livelihood outcomes arise from gender differences in agricultural commercialization.

An important livelihood outcome to which agricultural commercialization has been linked both positively and negatively is food and nutrition

DOI: 10.4324/9781003202004-4

security (von Braun and Kennedy, 1994, Carletto et al., 2017). Resource (particularly land, labour, and capital) allocation and policy decisions related to food and non-food crop production at all levels have implications for food supply and food prices, which affect food availability and access (Timmer, 2010). This is exemplified by what has become known as the 'world food crisis of 2007–2008', which some have partly attributed to resource allocation to biofuel production (Deininger and Byerlee, 2011, Koizumi, 2015).

The foregoing relates to longstanding debates about whether or not, and how, resource allocation towards the production of non-food cash crops could affect food security. Some have argued that large-scale land acquisitions for non-food cash crops, including the production of biofuels, compromise food security (Deininger and Byerlee, 2011, Koizumi, 2015). Similarly, for smallholder agricultural households, resource allocation between non-food cash crop and food crop production could have implications for food security (Govereh and Jayne, 2003, Dzanku and Sarpong, 2011). This implies that the association between agricultural commercialization and food security could depend on crop choice because, whereas staple food crops such as maize, cassava, plantains, and rice often play the dual role of food and cash crops, non-food cash crops such as cocoa, rubber, or even oil palm are produced exclusively as cash crops.

While some studies have examined gender differences in household food security (Kassie et al., 2014, Mason et al., 2015) and commercialization (Njuki et al., 2011, Fischer and Qaim, 2012), there is a dearth of literature on whether the relationship between agricultural commercialization and food security is gendered. Besides, certain questions remain unexplored. For example, the roles of commercialization capacity and pathways – defined as control over land and crop choice – as well as the relationship between gendered commercialization and specific measurements of food security, are not clearly delineated. This chapter makes a contribution to addressing these gaps in the literature in two ways. First, it examines gender differences in commercialization capacity with a focus on land access at the level of plots, tenure security, crop choice, and intra-household gender differences in the levels of commercialization. Second, it analyses gender differences in the association between agricultural commercialization and food security, using different food security measurements. We argue that if commercialization is gendered, as the literature (von Braun and Kennedy, 1994, Quisumbing et al., 2015) suggests, then there is reason to hypothesize that the association between commercialization, commercialization capacity, and food security are also likely to be gendered.

The rest of the chapter is organized as follows: the next section presents an overview of the existing relevant literature; this is followed by a discussion of the study context and an account of the methodology, including a description of the data and the analytical approaches employed. The penultimate section presents and discusses the results. The final section is the conclusion.

Agricultural commercialization and food security: a framing

Agricultural commercialization can be defined at various levels: national, sub-national, and household. In this study, we focus on household-level commercialization where the phenomenon refers to the process of increasing the proportion of agricultural output sold by farmers (Pradhan et al., 2015).

Increased agricultural commercialization is viewed as essential for improving the income and living standards of smallholder farmers (Braimoh, 2009). This suggests a positive association between agricultural commercialization and smallholder livelihood outcomes, including food security. However, the relationship between commercialization and food security can be complicated – there are important spatial specificities as well as gender and class differences in the ways people experience the processes and outcomes of food security (Dzanku et al., 2021).

Based on the literature, Ruel and Alderman (2013) identified four pathways through which agricultural production could influence food security and nutrition: food prices, agricultural income, self-provisioning, and gender issues. These pathways form the conceptual foundation for analysing the relationship between commercialization and food security.

The association between commercialization and food security could also depend on the dimension of food security (availability, access, utilization, stability) that is measured. For example, using a small sample of households in the Ashanti Region of Ghana, Anderman et al. (2014) found that, while resource allocation to cocoa and oil palm production had a significant negative effect on food availability, there was no significant association with household dietary diversity (HDD). Yet, others (Pierre-Louis et al., 2007, Komarek, 2010, Negash and Swinnen, 2013) have found a significant positive effect of commercialization on both food availability and nutrition.

Some important missing pieces of these studies are whether the hypothesized effects differ by commercialization pathways (i.e., commercialization through the generation of food crop surpluses or differing capacity to produce non-food cash crops), and by gender, which is what this chapter seeks to contribute to the literature.

The gendered nature of agricultural commercialization and food security

Agricultural commercialization is often associated with both absolute and relative decline in women's control over resources because men dominate the cash crop sector (Fischer and Qaim, 2012). Though women make essential contributions to agriculture and rural livelihoods, their access to productive resources, such as land and capital, is often constrained, which limits their capacity to venture into commercial agriculture (FAO et al., 2011). As noted by Fischer and Qaim (2012), women are increasingly disadvantaged as overall rates of commercialization increase because of persistent gender disparities in access to productive resources.

Commercialization is often associated with the adoption of sophisticated and new technologies, which may further lessen the role of women (Fischer and Qaim, 2012). In our research sites, for instance, we observed that the few bullock ploughs available were used by men and became available for women's use only when men had finished ploughing their fields (Tsikata and Eweh, Forthcoming). Due to these forms of inequities, some empirical studies in rural sub-Saharan Africa (SSA) have found that female-headed households (FHHs) are more vulnerable to food insecurity than male-headed households (MHHs) (Kassie et al., 2014, Mason et al., 2015) and that the gap is mainly related to inequalities in the control of productive resources (mainly land and labour).

The above notwithstanding, because men and women spend money differently, with women having a higher propensity to spend the income they have control over on household necessities such as food (Meinzen-Dick et al., 2012), it is possible that commercialization could enhance women's food security more than men's even if women earned less from such market participation. Indeed, there is evidence that household food and nutrition security is better enhanced when women have more control over household income (Quisumbing et al., 1995, Sinclair et al., 2016).

The study context

Land and agricultural commercialization has a longstanding history in Ghana from the colonial period when first rubber, then oil palm, and later cocoa were introduced as export commodities in southern Ghana, and food was grown in northern and southern Ghana to feed local and regional markets. Consequently, farmers have increasingly become less self-sufficient in food production and have purchased an increasing proportion of their food needs (Shepherd, 1981). These processes accelerated in the post-colonial period with increasing urbanization and the expansion of the informal trade in locally grown and imported fresh and processed food, and with these processes, the increasing diversification of agrarian livelihoods.

In the early independence period, state policies were generally biased in favour of large-scale commercial agriculture. Direct state interventions in commercial agricultural activities culminated in the monopolization of the domestic cocoa purchasing market by the Cocoa Marketing Board in 1961. State-owned agricultural corporations were also established, leading to further marginalization of smallholder farmers (Killick, 1978). The seven-year development plan (1964–1970) established direct control of food crop commercial agriculture through the establishment of the State Farms Corporation.

In the 1980s, economic liberalization led to the deregulation of factor input and output markets, which attenuated direct state interventions in agriculture. This was short-lived, however, as the 1982–1983 food shortages fuelled renewed policy attention towards self-sufficiency as a safeguard against food shortages and price volatility.

It can be argued that a coherent policy framework for supporting smallholder agriculture with the aim of reducing food insecurity started in the 1990s (Dzanku and Udry, 2017), but cocoa continued to be prioritized, although some deregulation of internal cocoa marketing was implemented. The 1990s also marked the beginning of policy attention towards non-traditional export crops, with the aim of diversifying Ghana's agricultural exports.

In the 2000s, Ghana experienced the great global land rush that saw large tracts of land acquired by transnational corporations and their Ghanaian partners for growing biofuels, fruits, vegetables, and maize and also for speculation. Studies in this period established the gendered character of labour regimes of large-scale land acquisitions, be it contract farming or plantation agriculture and their propensity to deepen rural differentiation and land dispossession (Boamah, 2014, Tsikata and Yaro, 2014).

Commercial agriculture in Ghana today operates under four distinct but related models: smallholder agriculture, medium-scale independent farms, plantation agriculture, and contract farming (Yaro et al., 2017). The smallholder model, which is based on a household-based production system, remains the most dominant both in terms of aggregate scale and number of participants, making it the most important in terms of potential impact on livelihood outcomes, including food security. Not only are these models distinct in their utilization of productive resources, but they are also associated with varying degrees of social stratification including along the lines of gender and class (Hall et al., 2017). Gender differentiation stems from gender-based gaps in access and control of land, labour, and capital as well as biases in recruitment practices in respect of contract farming arrangements (Torvikey et al., 2017). Although the smallholder model is relatively more gender inclusive, other forms of difference exist, including inequalities in access to land, labour, and other productive resources (Agbosu et al., 2007, Tsikata, 2016). In spite of its dominance, the smallholder model is also associated with poorer livelihood outcomes – both in terms of consumption poverty and food insecurity (Ghana Statistical Service, 2018) – which some (Tsikata and Yaro, 2014) argue results from decades of policy neglect.

Important spatial variations exist within the overall country context and the context of the present study areas with respect to the three main themes of this chapter: commercialization, food security, and gender. From the past to the present, commercial agriculture in southern Ghana has been dominated mainly by Ghana's traditional agricultural export commodities (cocoa, oil palm, and rubber). Per contra, commercial agriculture in northern Ghana thrives mainly on the production of food crops (cereals, legumes, yam, and some vegetables). This north–south divide is conditioned by several factors including agroecology (a unimodal rainfall pattern and fragile soils in the north compared to a bimodal rainfall pattern and less fragile soils in the south). There is also the long history of commercial agriculture concentration in the south fuelled by state policies that prioritized tree crop export commodity agriculture to the neglect of food crops. These have combined to create poorer livelihood outcomes in the north than in the south, including longer spells of food insecurity in the absence of irrigation agriculture (Ghana Statistical Service 2018). There are also north–south differences in sociocultural norms and descent rules that create wider gender-based livelihood capital gaps in the north than in the south (Agbosu et al., 2007, Apusigah, 2009). Our study areas were chosen to reflect the above differences and similarities.

Methodology

Data

This chapter uses cross-sectional household survey data collected in 2020 from 14 communities located in four districts of the then Eastern, Brong-Ahafo, Northern, and Upper East regions of Ghana.[1] The districts are Kwaebibirem, Asunafo North, East Gonja, and Garu-Tempane. These districts and the communities were purposively chosen based on the different types/models of commercialization identified by the research team during a reconnaissance survey in 2015. Kwaebibirem was chosen for oil palm and the presence of large- and medium-scale oil palm-related agro companies. Cocoa is also a major cash crop in Kwaebibirem but with relatively younger farms compared with Asunafo North that was chosen mainly for its cocoa production. East Gonja and Garu-Tempane were chosen for food crop commercialization – yam and rice for East Gonja, and sorghum and vegetables for Garu-Tempane.

The aim was to achieve a statistically representative sample for each community based on a sampling frame generated through a census of all households in the selected communities. Based on a probability proportional to size sampling method, 769 households were randomly drawn. Approximately 25% of the households are female headed. Our units of analysis are the household and plot levels. The plot-level data contains 1,799 plots cultivated by 755 households.

Analytical methods

We first describe the main indicators used for achieving the objectives of the study, followed by a description of the empirical multivariate analytical methods.

Key indicators

The main variables of interest were measures of agricultural commercialization, gender, and food security. For agricultural commercialization, we focused on crops and followed Strasberg et al. (1999) to construct a crop commercialization index (CCI) as:

CCI = 100 × [gross value of crop sales/gross value of all crops produced].

CCI equals zero for pure subsistence households and one for purely commercial households. We also used the gross value of crop sales as a commercialization measure because the CCI could be driven by distress sales (Dzanku et al., 2021).

Our inter-household gender analyses compare FHHs with MHHs. While 'head' in this sense connotes control over resources and decision-making within the household, it is often a more complex concept embedded in sociocultural and religious norms about who or what 'head' means or should be. In this study, we accepted whoever the household identified as head, and in the study context, this was often the husband or male in the case of conjugal or couple households irrespective of who controlled resources or had the most decision-making power. Female headship was largely restricted to non-couple households or other pathways to female headship, such as separation, divorce, widowhood, and long absence (six months or more) of a male who would otherwise be identified as head.

The headship-based analysis is limiting because, for example, females in MHHs and females in FHHs face different constraints (Doss and Morris, 2001). Also, FHHs are heterogeneous with 'variations in women's routes to headship' (Chant, 2004, p. 21). Therefore, beyond the gendered headship-based analyses, we also analysed gender differences focusing on plots operated by men and women separately. We collected plot-level gender-disaggregated data using the concept of plot holder or 'owner', which indicated some level of control over a plot. 'Owner' here is not necessarily in the *de jure* sense but means that the plot holder can make short- to medium-term production decisions, including the disposal of output from the plot.

Our main outcome variable is food security. We employed two measures at the household level to capture aspects of the dimensionalities of food security. This is important because agricultural commercialization could have different effects on food security depending on which dimension is examined. The first measure of food security is food availability throughout the year. We use responses to the question about whether the household had 'sufficient' food, experienced 'some shortage', or experienced 'extreme shortage' during each month of the year to construct a food adequacy score (FAS) as a measure of household food security. This measure addresses the food availability dimension of household food security. A score of one was assigned if a household experienced 'extreme shortage', two for 'some shortage', and three for 'sufficient' food. These scores were summed up to give an indicator of annual food adequacy – the higher the score, the more food secure a household was. With the maximum possible score of 36, we divided the annual FAS by 36 and multiplied by 100 to get a percentage point interpretation.

The second food security indicator measured HDD following FAO guidelines (FAO, 2010) as based on 12 food groups (cereals, roots and tubers, vegetables, fruits, meat, poultry, offal, eggs, fish and seafood, pulses/legumes/nuts, milk and milk products, oil/fats, sugar/honey, and miscellaneous). HDD ranges between 0 and 12. However, to make the indicator more intuitive to interpret, the score is divided by 12 and multiplied by 100 to give the percentage of total number of food groups from which a household consumes. It must be noted the HDD has limitations, including the fact that it does not measure the quantities and proportions of the household's food basket represented by the food groups. HDD also assumes that all types of food are equally important to households, ignoring issues of food culture.

Other variables

In analysing the determinants of household food security, we used key variables identified by Scholes and Biggs (2004) in their review of over 500 citations, which show that the most frequently cited determinants of household food (in)security are: poverty, education, food prices, employment, property rights, market access, and climate/environmental factors. The choice of variables to include in our econometric models is guided by these determinants. First, we include household and household-head characteristics: age, household size, child dependency ratio, sex ratio, marital status, and literacy rate. Second, we use variables that capture household economic and wealth status: availability of off-farm income opportunities, durable assets, livestock wealth, and total cultivated area, all of which are expected to be positively correlated with food security. Third, because households need not rely on their own resources to meet their food needs, we also included two 'network capital' variables that we expect to enhance household food security: remittances received and social-economic network capital, for which we used group memberships and associations as proxy. Fourth, the number of food crops produced by the household is included to capture household food crop diversity, which is expected to be positively associated with food security (Di Falco and Chavas, 2009).

Finally, shocks – both covariate and idiosyncratic – are known to be major threats to food security (Godfray et al., 2010, Akter and Basher, 2014). Therefore, we included in the food insecurity equation two variables that sought to capture shocks: (a) crop failure during the previous season and (b) the number of other shocks (e.g., death of income-earning member, loss of livestock, input price and output price shocks, etc.) suffered by the household. Village fixed effects were included in all the regressions to capture other covariate unobserved shocks and agroecology effects.

Regression models

We addressed the first objective of gender differences in commercialization capacity by analysing plot-level gender differences in access to land, tenure security, and type of crop produced. For the gender difference in land access and tenure security, we estimated the following regression models:

$$\ln(PlotSize_{ij}) = \alpha_1 + \delta_1 PlotControl_{ij} + \pi_1' X_{ij} + \beta_1' X_i + \varepsilon_i \tag{2.1}$$

$$PlotSecurity_{ij} = \alpha_1 + \delta_1 PlotControl_{ij} + \pi_1' X_{ij} + \beta_1' X_i + \varepsilon_i \tag{2.2}$$

where Log(*PlotSize*) and *PlotSecurity* are the log of plot size and plot holder's perception of the security of plot tenure, *PlotControl* measures the gendered access to land variable, X_{ij} and X_i are the vectors of other plot and household-level variables that influence plot size, and ε is the error term that captures unobserved shocks. The perceived plot tenure security indicator is a binary variable taking on the value one if the plot holder indicates that they have full control over the plot (i.e. they can plant or change current crops including tree crops) and have no fear of losing the plot. If a plot holder claims to have full control over their plot but has fears of losing it, then they are considered to have insecure tenure. The *PlotControl* variable is the indicator that captures the gender of the plot holder, and it takes on the value one if female and zero if male.

Still on the first objective where we consider the type of crop produced as indicative of the capacity to commercialize, we specify the following binary (2.3) and semi-continuous (2.4) equations for the decision of a farmer in household *i* to cultivate crop *j* on plot *k* as:

$$Crop^{\star}_{ijk} = \alpha_2 + \delta_2 PlotControl_{ijk} + \pi_2 X_{ik} + \beta_2' X_i + \varepsilon_{ijk},\ Crop_{ijk} = 1[Crop_{ijk} > 0], \tag{2.3}$$

and

$$\begin{aligned} Crop^{\star}_{ijk} &= \alpha_2 + \delta_2 PlotControl_{ijk} + \pi_2 X_{ik} + \beta_2' X_i + \varepsilon_{ijk}; \\ &\quad \text{where } \varepsilon_{ijk} \sim \text{Normal}(0, \sigma_\varepsilon^2) \\ Crop_{ijk} &= \max(Crop^{\star}_{ijk}, 0) \end{aligned} \tag{2.4}$$

We use probit and Tobit estimators for equations (2.3) and (2.4), respectively.

Our second objective concerns the association between food security and commercialization, and the associated heterogeneities in the relationship. Here, we specify the following linear regression model:

$$FS_i = \alpha_3 + \delta_3 CCI_i + \beta_3' X_i + \varepsilon_{3i} \tag{2.5}$$

where *FS* is either the annual FAS or HDD indicators, CCI is the commercialization indicator (defined earlier); the βs are unknown parameters associated with the vector of exogenous control variables *X*, which includes the gender of household head and location dummies; and ε_i is the random error term. We use the interactions approach to test the hypothesis that the association between commercialization and food security is the same for the genders by specifying:

$$FS_i = \alpha_0 + \lambda_0 FHH_i + \delta CCI_i + \lambda_1 (FHH_i \times CCI)_i + \beta' X_i + \varepsilon_i \tag{2.6}$$

The first hypothesis of interest, which is that the returns to commercialization (in terms of food security) are the same for FHHs and MHHs, is given by $\lambda_1 = 0$. Another hypothesis of interest is that, on average, the association between commercialization and food security is identical for FHHs and MHHs at the same commercialization rate, that is, $\lambda_0 = 0, \lambda_1 = 0$. This hypothesis answers the question: if FHHs and MHHs achieve the same level of commercialization, would their level of food security be the same?

Results

Descriptive statistics

Table 2.1 provides summary statistics of key household-level variables. We observed a large difference in the gross value of crop sales between FHHs and MHHs in both the northern and southern samples: the gross value of

Table 2.1 Descriptive summary statistics, by sex of household head

	North			*South*		
	FHH	*MHH*	*Diff.*	*FHH*	*MHH*	*Diff.*
Variables	*n =* *79*	*n =* *345*	*p value*	*n =* *113*	*n =* *232*	*p-val.*
Gross value of crop sales (US$100)	0.69	2.99	0.00	7.32	18.75	0.00
Crop commercialization index (%)	29.28	30.83	0.64	80.08	83.10	0.23
Food adequacy (%)	81.40	87.41	0.00	91.18	93.85	0.00
Dietary diversity (%)	25.53	31.76	0.00	38.42	38.90	0.76
Age of household head	59.28	51.30	0.00	52.78	53.03	0.88
Household size	5.32	6.90	0.00	3.42	4.64	0.00
Under-5 dependency ratio	0.25	0.28	0.55	0.24	0.27	0.62
Sex ratio	1.48	1.20	0.02	1.29	1.11	0.07
Married head	0.20	0.94	0.00	0.24	0.89	0.00
Years of schooling	0.43	3.41	0.00	6.55	9.78	0.00
Farm size (hectare)	3.21	8.21	0.00	7.32	14.61	0.00
Received off-farm income	0.78	0.78	0.92	0.93	0.92	0.72
Off-farm income (US$100)	2.66	3.56	0.35	8.67	12.95	0.00
Normalized asset index	0.12	0.17	0.00	0.27	0.35	0.00
Livestock wealth (cow equivalent)	1.22	2.38	0.02	0.92	1.85	0.00
Received remittance	0.16	0.10	0.08	0.49	0.36	0.03
Farmer association member	0.09	0.10	0.79	0.20	0.66	0.00
Credit association member	0.62	0.57	0.40	0.06	0.06	0.92
Religious group member	0.25	0.28	0.62	0.78	0.69	0.08
Political party member	0.04	0.21	0.00	0.07	0.08	0.72
Number of staple foods produced	2.14	2.86	0.00	1.14	1.34	0.10
Agricultural production-related shocks	0.68	0.68	0.97	0.20	0.28	0.11
Household-specific non-crop shocks	0.76	0.76	0.96	0.45	0.38	0.20
Land title certificate	0.01	0.02	0.77	0.21	0.43	0.00

Source: Authors, using household survey data

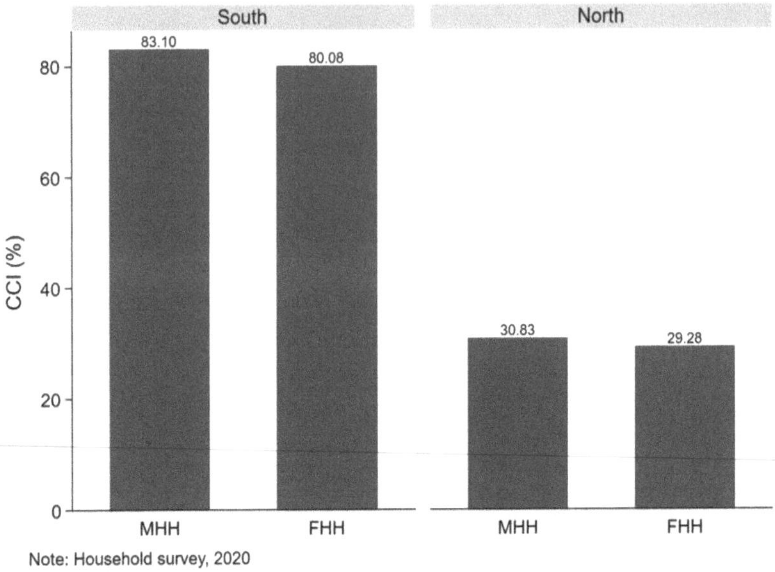

Figure 2.1 Gender and agroecology differences in crop commercialization

sales among FHHs is only 23% and 39% of the sales of MHHs in the north and south, respectively. The CCI values indicate that approximately 31% and 82% of the gross value of crops produced by the north and south households, respectively, were sold. The large north–south gap in CCI (Figure 2.1) was driven by the exclusive production of cocoa and oil palm – two of Ghana's leading export crops – in the southern study areas.

In spite of the large gender gap in the gross value of crop sales, there was no gender gap in CCI. This could be indicative of distress sales, which have been found to be more common among FHHs (Dzanku et al., 2021). Other studies using data from Malawi, Tanzania, and Uganda also show that although FHHs had a lower rate of market participation, their CCI conditional on participation was higher than that for MHHs (Carletto et al., 2017).

We observed that food was significantly more available throughout the year in MHHs than in FHHs in both the northern and southern zones, but while diets were of identical diversity between MHHs and FHHs in the south, MHHs had a significantly more diverse diet in the north. So either way, FHHs in the north were less food secure than MHHs.

Beyond the gendered headship-based analysis, we observed that a large share of plots (about 70%) were controlled by men, suggesting that women's capacity to commercialize was lower than men's (Figure 2.2).

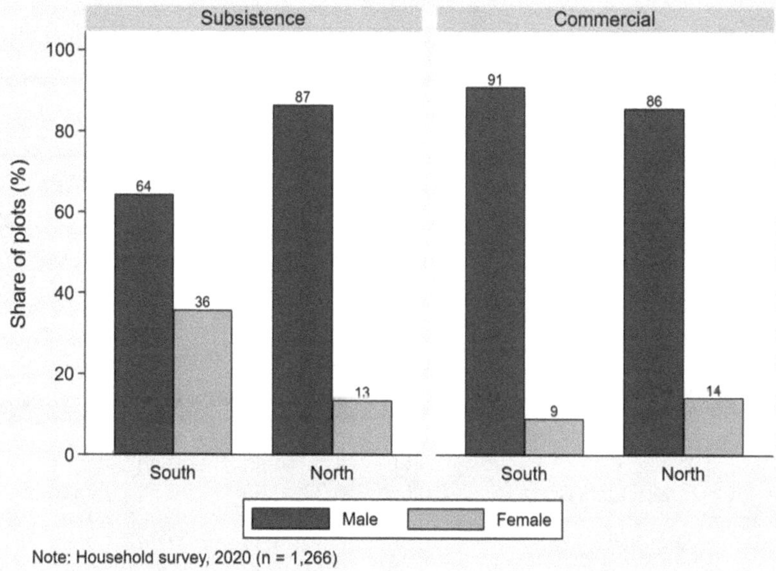

Figure 2.2 Land ownership and control within couple households, by ecological zone and market orientation

Table 2.2 presents plot-level mean summary statistics of key variables. We observed more statistically significant (at the 0.05 level) plot-level gender differences in the north than in the south. Female farmers had significantly fewer claims over resources that enhance the capacity to commercialize. For example, they held only 59% and 73% of the land sizes held by men in the north and south, respectively. However, women did not feel significantly less secure on the plots they operated on. The sources of land could shed more light on this result. Also, people's sense of security is not necessarily objective. As the literature suggests, certain terms and conditions of land access, including the interests women acquire when using land belonging to their husbands, are generally less insecure than the usufructuary interest (Agbosu et al., 2007).

On the type of crop produced, we observed in the north that vegetables (okro, pepper, and onions) and pulses (groundnuts and soybean) were significantly more commonly produced by women while roots and tubers (mainly yam) were produced mainly by men. Maize, cassava, and plantains were the main staple foods in the southern study areas, and whereas there were no plot-level gender differences in grains production, cassava and plantains were significantly more commonly found on women's plots in the south. Participation in tree crops (mainly cocoa and oil palm) production, which define commercial agriculture in the south, did not appear gendered at the plot level

Table 2.2 Plot-level summary statistics, by sex of plot holder

	North			South		
	Female	*Male*	*Diff.*	*Female*	*Male*	*Diff.*
Variables	*n = 279*	*n = 632*	*p value*	*n = 254*	*n = 634*	*p-val.*
Plot size (acre)	2.05	3.46	0.00	3.46	4.70	0.00
Tenure security dummy	0.72	0.87	0.00	0.88	0.84	0.13
Hired-in labour	0.38	0.35	0.43	0.78	0.76	0.70
Hired labour (US$/acre)	9.55	7.46	0.08	31.09	31.98	0.76
Used exchange labour	0.29	0.31	0.66	0.13	0.10	0.35
Exchange labour (person-days/acre)	3.67	3.15	0.39	1.42	1.02	0.43
Male family labour (person-days/acre)	10.96	19.23	0.00	13.03	31.46	0.00
Female family labour (person-days/acre)	19.04	13.50	0.00	39.67	14.38	0.00
Irrigated plot	0.11	0.14	0.38	0.00	0.01	0.67
Roots, tubers, and plantains	0.05	0.22	0.00	0.36	0.26	0.00
Vegetables	0.22	0.07	0.00	0.00	0.01	0.40
Grains	0.66	0.69	0.36	0.06	0.08	0.25
Pulses	0.38	0.24	0.00	n.a.	n.a.	n.a.
Tree crops	0.00	0.01	0.46	0.88	0.88	0.88
Gross value of crops produced (US$)	164.65	376.60	0.00	499.92	854.79	0.00
Gross value of crop sales (US$)	60.70	152.14	0.01	436.07	731.11	0.00
Crop commercialization index (%)	30.42	29.30	0.63	80.10	78.92	0.58

Source: Authors, using household survey data

(Table 2.2), although there were significant gender differences in the scale of production.

Lastly, we observed that, as with the household headship-based analysis, the gross value of crop sales from women's plots was significantly lower than that from men's plots, but we observed no difference in the plot-level commercialization index, showing that how commercialization is measured matters for conclusions one might reach.

Next, we turn to the regression analysis where we provide further insight after controlling for other confounding factors.

Regression results and discussions

The gendered nature of commercialization capacity

We began regression analyses by estimating equations (1) and (2), which examine plot-level gender differences in access to land and the associated perception of tenure security (Table 2.3). In both the north and south, the coefficients of the female plot variable carried the expected negative sign and were both statistically significant and large (columns 1 and 2). On average, women's plots were approximately 21% and 37% smaller than men's in the south and north, respectively. One could attribute the larger gender gap in the north to more favourable lineage descent rules and social norms in the south, as well as to practices such as women using men's fallow plots to farm in Gonja-East, and the high population density in Garu-Tempane (Tsikata and Eweh, forthcoming).

Columns (3) and (4) show that while the female plot holder coefficients had the expected negative signs in both zones, they were not statistically significant at any conventional level. This means that women and men were equally likely to report that they had full control over the plots they

Table 2.3 Gender and commercialization capacity: plot size and tenure security

	(1)	*(2)*	*(3)*	*(4)*
Variables	*Plot size (South)*	*Plot size (North)*	*Tenure (South)*	*Tenure (North)*
Female plot holder	−0.212***	−0.370***	−0.002	−0.054
	(0.071)	(0.067)	(0.025)	(0.040)
Access to lineage land	−0.060	0.349***	0.326***	0.213***
	(0.054)	(0.058)	(0.033)	(0.039)
Exchange labour available	0.076	0.065		
	(0.079)	(0.056)		
Adult male labour	0.040	0.016		
	(0.037)	(0.022)		
Adult female labour	0.044	0.017		
	(0.029)	(0.025)		
Number of dependants	0.002	0.047***		
	(0.025)	(0.014)		
Age of head/100	0.722***	−0.136	0.553***	0.015
	(0.248)	(0.301)	(0.131)	(0.145)
Age of spouse/100	−0.176	0.705*	−0.262	0.035
	(0.307)	(0.391)	(0.160)	(0.204)
Head's years of schooling	0.003	−0.011*	0.002	0.001
	(0.007)	(0.006)	(0.003)	(0.004)
Spouse's years of schooling	0.008	0.027***	−0.006**	0.007
	(0.007)	(0.010)	(0.003)	(0.006)
Normalized asset index	0.799***	0.784**	0.199**	−0.748***
	(0.233)	(0.341)	(0.098)	(0.205)
Log off-farm income	−0.010	−0.015		
	(0.015)	(0.012)		
Migrant	−0.048	−0.179**	−0.004	−0.091
	(0.066)	(0.089)	(0.029)	(0.058)
Religious group member	0.026	0.121*	0.059*	0.005
	(0.070)	(0.072)	(0.031)	(0.042)
Constant	0.548**	0.379***	0.337***	0.952***
	(0.236)	(0.143)	(0.086)	(0.076)
District fixed effects	Yes	Yes	Yes	Yes
R-squared	0.118	0.237	0.222	0.167
Observations	888	911	888	911

Standard errors in parentheses; * $p < 0.10$, ** $p < 0.05$, *** $p < 0.01$

cultivated. It seems thus far that the availability of land is the main driver of the gender differences in commercialization capacity. Another potential determinant is crop choice or the lack thereof, and it is to this that we now turn.

Tables 2.4 and 2.5 report the plot-level crop choice regression results. Commercialization in the south is predominantly driven by cocoa and oil palm. After adjusting for other covariates using regression analysis, however, we found no evidence that crop choice was gendered in the southern study areas (Table 2.4). There are three key drivers of cocoa production in the sample: the ratio of the land to labour endowment, adult male labour, and input subsidy. In the case of oil palm production, the drivers are the land–labour ratio, female adults, and the availability of contract farming arrangements. The finding that resource allocation to both cocoa and oil palm increased with higher land relative to household labour suggests that the production

Table 2.4 Gender and commercialization capacity: area cultivated with crops in the south

	(1)	*(2)*	*(3)*
Variables	*Cocoa area*	*Oil palm area*	*Staples area*
Female plot holder	0.318	−0.301	0.200
	(0.289)	(0.213)	(0.274)
Land–labour ratio	0.139***	0.051**	−0.022
	(0.038)	(0.021)	(0.025)
Log distance to plot	0.262	0.172	0.059
	(0.172)	(0.124)	(0.143)
Adult males	0.276**	0.079	−0.076
	(0.120)	(0.111)	(0.151)
Adult females	0.021	0.134*	0.044
	(0.158)	(0.081)	(0.100)
Received input subsidy	0.831***	−0.338*	0.447**
	(0.234)	(0.174)	(0.208)
Log off-farm income	−0.005	0.026	−0.022
	(0.050)	(0.034)	(0.046)
Livestock per adult	−0.257	−0.041	−0.017
	(0.168)	(0.097)	(0.164)
Age of head/100	0.774	−0.549	−2.205***
	(1.117)	(0.894)	(0.840)
Age of spouse/100	−1.217	0.862	1.254
	(1.621)	(1.121)	(1.210)
Head's years of schooling	0.018	0.006	0.003
	(0.028)	(0.022)	(0.024)
Spouse's years of schooling	0.035	0.031	−0.018
	(0.028)	(0.027)	(0.025)
Contract farmer		1.147***	
		(0.168)	
Village fixed effects	Yes	Yes	Yes
Observations	888	888	888

Standard errors in parentheses; * $p < 0.10$, ** $p < 0.05$, *** $p < 0.01$

Table 2.5 Gender and commercialization capacity: area cultivated with crops in the north

	(1)	*(2)*	*(3)*	*(4)*
Variables	*Yam area*	*Grains area*	*Vegetables area*	*Pulses area*
Female plot holder	−1.788***	−0.779***	0.366***	0.493***
	(0.285)	(0.177)	(0.067)	(0.125)
Land–labour ratio	0.000	0.183***	0.011	0.064**
	(0.030)	(0.040)	(0.014)	(0.030)
Log distance to plot	0.204	−0.136**	0.022	−0.048
	(0.138)	(0.056)	(0.027)	(0.055)
Adult males	−0.094*	0.143*	0.074***	0.060
	(0.048)	(0.084)	(0.028)	(0.054)
Adult females	0.069	0.174***	−0.032	−0.022
	(0.059)	(0.057)	(0.029)	(0.053)
Received input subsidy	0.001	0.010	−0.123***	−0.283*
	(0.209)	(0.194)	(0.116)	(0.161)
Log off-farm income	0.007	−0.022	0.033**	0.047
	(0.029)	(0.036)	(0.015)	(0.029)
Livestock per adult	−0.071	0.164**	−0.006	0.076
	(0.083)	(0.081)	(0.050)	(0.061)
Age of head/100	−0.053	−0.083	−0.200	−0.783
	(0.966)	(0.621)	(0.306)	(0.552)
Age of spouse/100	2.125*	0.951	1.085***	2.068***
	(1.291)	(0.907)	(0.386)	(0.778)
Head's years of schooling	−0.000	−0.025*	0.003	−0.013
	(0.019)	(0.014)	(0.008)	(0.013)
Spouse's years of schooling	0.015	0.026	0.002	0.044**
	(0.025)	(0.028)	(0.012)	(0.021)
Village fixed effects	Yes	Yes	Yes	Yes
Observations	470	911	911	911

Standard errors in parentheses; * $p < 0.10$, ** $p < 0.05$, *** $p < 0.01$.

of these crops is relatively more constrained by land than labour per se. This finding, however, needs to be qualified because it does not account for the use of hired labour and the widespread use of weedicides in the farming system, a practice attributed to labour shortages and affordability (Tsikata and Eweh, forthcoming).

We found strong gender differences in the type of crop produced at the plot level in the northern sample (Table 2.5). Yam is the main cash crop in the northern district of East Gonja, and female plot holders allocated significantly less land to the crop: about 1.8 acres less, on average, which was a large gender gap given that the unconditional average area allocated to yam in the sample was 1.4 acres. Grains, particularly rice and maize, are also important cash crops in the two northern districts, and here too, females allocated significantly less land to the crops (nearly 0.8 acres less, on average). Pulses (mainly groundnuts and beans) and vegetables are grown equally for food and cash, and we found that women had significantly more land allocated to these crops: women devoted 0.36 and 0.49 acres more land to vegetables and pulses. These findings confirm our qualitative studies, which establish that women's crop choices were linked with their responsibilities for household food provisioning. This is an important issue for household food security because the crops produced by men were considered to be the real food and women's crops were often referred to as condiments.

The southern Ghana findings confirm the evidence from earlier qualitative studies that, in some cases, women do not specialize in specific crops (Doss, 2002, Lambrecht et al., 2018).

Gender, commercialization, and food security

Econometrically, it is difficult to identify the effect of commercialization on food security because it can be argued that commercialization is determined simultaneously with food security. Secondly, it is possible that the same unobserved factors that determine households' commercialization decisions are also correlated with food security. Both issues raise endogeneity concerns when estimating the relationship between commercialization and food security, particularly in small samples. In the absence of a convincing instrumental variable (i.e., a variable that is highly correlated with commercialization but influences food security only through its effect on commercialization), the results presented here should be interpreted as correlations, not causal effects.

Tables 2.6 and 2.7 present results from analysing the association between agricultural commercialization and food security and provide tests for gender differences as well as the contribution of particular food security measurements in the food security–commercialization relationship. We found that commercialization had a statistically significant but small positive association with food adequacy in the south – a 1% increase in CCI is associated with about 0.05% increase in access to adequate food throughout the year (Table 2.6, column 1). The very small magnitude of association was partly because reported food adequacy levels were already very high, about 94% on average, in the south. We did not find evidence of either a statistically significant or practically meaningful association between commercialization and food adequacy in the northern sample (Table 2.7). This shows that commercialization is not a panacea to food insecurity – it has been shown that some households can appear highly commercialized when in fact their market participation does not mean accumulation but rather 'distress push commercialization' (Dzanku et al., 2021)

Does the choice of food security measurement matter in the food security–commercialization relationship? It does, to some extent, in the south, where we found a significant commercialization effect on food adequacy but not a significant effect on dietary diversity. Commercialization had no significant association with either food adequacy or diet diversity in the north, although the commercialization coefficients carry opposite signs in both food security equations, suggesting that the direction of association may not be the same, depending on what dimension of food security one measures. A larger sample may be required to shed more light on this.

Table 2.6 Commercialization and food security: southern study areas

Variable	*(1)* *Food adequacy*	*(2)* *Food adequacy*	*(3)* *Dietary diversity*	*(4)* *Dietary diversity*
CCI	0.047**	0.044*	0.030	0.009
	(0.023)	(0.024)	(0.031)	(0.036)
Female-headed household	−2.940**	−3.814	4.758**	−0.233
	(1.283)	(4.130)	(2.178)	(4.434)
Female*CCI		0.011		0.063
		(0.049)		(0.053)
Age of household head	0.017	0.017	−0.042	−0.041
	(0.031)	(0.031)	(0.059)	(0.058)
Years of schooling	−0.004	−0.003	−0.140	−0.133
	(0.108)	(0.108)	(0.160)	(0.161)
Household size	−0.219	−0.220	−0.431	0.442
	(0.220)	(0.220)	(0.364)	(0.365)
Under-5 dependency ratio	−0.408	−0.400	−0.122	−0.078
	(1.030)	(1.034)	(1.821)	(1.824)
Sex ratio	0.844**	0.842**	−0.693	−0.706
	(0.408)	(0.409)	(0.820)	(0.825)
Couple household	−1.426	−1.410	2.289	2.383
	(1.187)	(1.194)	(2.182)	(2.199)
Log off-farm income	−0.122	−0.123	1.457***	1.450***
	(0.219)	(0.219)	(0.373)	(0.371)
Normalized asset index	8.312***	8.343***	16.481***	16.658***
	(2.275)	(2.277)	(4.352)	(4.354)
Livestock (cow equivalent)	0.062	0.063	0.054	0.058
	(0.197)	(0.197)	(0.428)	(0.429)
Log farm size	−0.117	−0.111	1.231	1.266
	(0.573)	(0.575)	(0.926)	(0.927)
Received remittance	0.971	0.984	−3.998**	−3.922**
	(0.847)	(0.850)	(1.588)	(1.601)
Political party member	0.991	0.962	−5.757**	−5.925**
	(1.348)	(1.353)	(2.901)	(2.921)
No. of staple foods produced	0.770*	0.771*	1.632**	1.638**
	(0.419)	(0.420)	(0.688)	(0.686)
Contract farmer	1.419	1.404	0.587	0.501
	(0.968)	(0.971)	(1.714)	(1.727)
Crop production shocks	0.326	0.307	2.951	2.846
	(1.301)	(1.315)	(1.942)	(1.944)
Non-crop related shocks	−1.240	−1.254	−0.733	−0.808
	(1.155)	(1.147)	(1.597)	(1.595)
Constant	85.241***	85.525***	22.785***	24.406***
	(3.367)	(3.379)	(5.081)	(5.376)
Village fixed effects	Yes	Yes	Yes	Yes
R-squared	0.133	0.133	0.223	0.225
Observations	345	345	345	345

Standard errors in parentheses; * $p < 0.10$, ** $p < 0.05$, *** $p < 0.01$.

Table 2.7 Commercialization and food security: northern study areas

	(1)	*(2)*	*(3)*	*(4)*
Variables	*Food adequacy*	*Food adequacy*	*Dietary diversity*	*Dietary diversity*
CCI	0.001 (0.021)	0.008 (0.024)	−0.035 (0.025)	−0.034 (0.029)
Female-headed household	0.647 (2.784)	1.495 (3.207)	−6.051* (3.081)	−5.924* (3.249)
Female*CCI		−0.031 (0.049)		−0.005 (0.053)
Age of household head	−0.046 (0.040)	−0.047 (0.039)	0.018 (0.048)	0.018 (0.048)
Years of schooling	0.207 (0.131)	0.207 (0.131)	0.061 (0.158)	0.061 (0.158)
Household size	−0.471** (0.225)	−0.474** (0.226)	0.195 (0.279)	0.195 (0.279)
Under-5 dependency ratio	−0.324 (1.554)	−0.312 (1.543)	−1.253 (1.912)	−1.251 (1.915)
Sex ratio	−0.394 (0.548)	−0.400 (0.555)	0.439 (0.733)	0.438 (0.733)
Couple household	−0.790 (2.318)	−0.864 (2.324)	−5.977** (2.918)	−5.988** (2.937)
Log off-farm income	0.195 (0.249)	0.192 (0.250)	0.646* (0.332)	0.646* (0.332)
Normalized asset index	4.887 (3.879)	4.834 (3.885)	17.183*** (5.942)	17.175*** (5.953)
Livestock (cow equivalent)	0.380*** (0.115)	0.379*** (0.117)	0.229 (0.195)	0.228 (0.195)
Log farm size	2.182** (1.003)	2.221** (1.006)	0.843 (0.926)	0.849 (0.928)
Received remittance	−4.185* (2.263)	−4.243* (2.275)	2.620 (2.198)	2.611 (2.211)
Political party member	−2.137* (1.294)	−2.161* (1.287)	6.341** (2.506)	6.338** (2.507)
No. of staple foods produced	−1.419** (0.700)	−1.450** (0.698)	1.166 (0.827)	1.161 (0.830)
Crop production shocks	2.974** (1.512)	2.950* (1.513)	1.739 (1.705)	1.736 (1.707)
Non-crop related shocks	−3.410** (1.645)	−3.454** (1.647)	−0.900 (2.066)	−0.907 (2.072)
Constant	75.163*** (4.930)	75.193*** (4.931)	18.944*** (4.070)	18.949*** (4.079)
Village fixed effects	Yes	Yes	Yes	Yes
R-squared	0.424	0.424	0.257	0.257
Observations	424	424	424	424

Standard errors in parentheses; * $p < 0.10$, ** $p < 0.05$, *** $p < 0.01$.

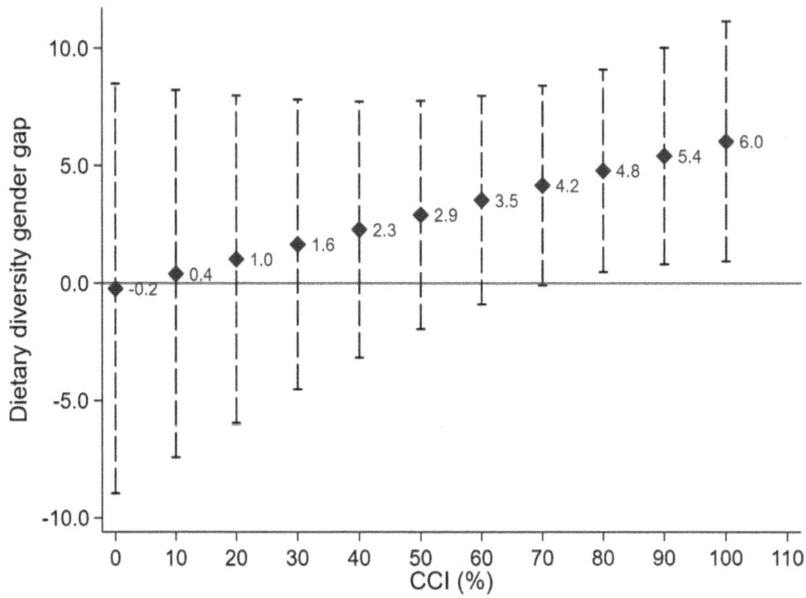

Figure 2.3 The gender gap in dietary diversity increases with commercialization
Note: The line caps represent 95% confidence intervals

Lastly, we ask two important gender questions regarding the association between food security and commercialization. First, is the association between food security and commercialization the same for FHHs and MHHs? The answer to this question is in the affirmative, irrespective of the agroecological context and food security measure employed (columns 2 and 4 of Tables 2.6 and 2.7).

Second, is the association between commercialization and food security identical for FHHs and MHHs at the same level of commercialization? Here, it depends on the agroecological context and how food security is measured. The answer is in the negative for both food security indicators in the southern sample, but largely in the affirmative in the northern study areas. For example, at median CCI of 89% in the southern sample, FHHs are estimated to have 2.8 percentage points lower food adequacy rate than MHHs ($p = 0.036$). However, at the same level of commercialization, FHHs have approximately 5.4 percentage points higher dietary diversity than MHHs ($p = 0.022$). Indeed, Figure 2.3 shows that dietary diversity is significantly higher among FHHs than MHHs at CCI greater than 70%. This result suggests that females have a higher tendency than males to prioritize dietary diversity when they control cash from agricultural commercialization (Ragasa et al., 2019).

The correlation between gendered commercialization and food security becomes clearer if we consider the survey results in light of the qualitative research findings in the study communities. First, not only were all the study communities experiencing seasonal food shortages, but there were also gender and class differences in the experiences of seasonal food shortages. Secondly, in one of the two most commercialized districts – Asunafo North – seasonal food insecurity was considered a major livelihood problem in spite of the high seasonal incomes from cocoa production. Thirdly, on the question of dietary diversity, the qualitative studies found a loss of certain foods as a result of the loss of the commons, the use of weedicides, and the use of fertilizer. Thus, while dietary diversity was considered a positive outcome of agricultural commercialization, particularly for FHHs, the indicator cannot account for the disappearance of certain important foods as a result of agricultural intensification, which is a consequence of commercial agriculture. Therefore, the survey finding of a somewhat positive association between commercialization and dietary diversity must be tempered with the opposing qualitative finding exposing the limitation of dietary diversity as a measure of food security in this context.

Conclusions

Agricultural policies in Africa aim at deepening agricultural commercialization and the drive towards large-scale production for the market. This is because commercialization is expected to improve the welfare of smallholders. This chapter has contributed to the longstanding debate about whether or not commercialization enhances household food security. First, it unpacks the gendered nature of commercialization processes by focusing on the factor endowments of land, labour, and crop production opportunities at the plot level. As expected, based on findings from the related literature, the lion's share of plots within the household is controlled by men; women farmers are relatively more land-constrained than men, and this tends to lower their capacity to commercialize across varying agroecological settings. On a more positive note, women do not feel less secure than men on the plots they cultivate. It is not clear whether this position is linked with the sources of the land they are cultivating. The qualitative literature suggests that, in spite of its growing scarcity, lineage or inherited land provides more tenure security than land given to women by their husbands or rented.

Still on the capacity to commercialize based on crop enterprise, except in relation to the differences in the size of their landholdings, women farmers did not appear disadvantaged in the more commercialized, matrilineal forest zones of the south, where no clear gender-based crop segmentation was evident, after adjusting for other crop choice correlates, particularly the ratio of land to labour and input subsidies. In the less commercialized, patrilineal, and drier northern zone, however, clear differences in crop enterprise-based commercialization pathways were identified, with women being less represented in relatively more lucrative pathways, particularly yam production. Women in the north tended to focus more on the crops that helped them satisfy their household food provisioning responsibilities, which is consistent with the literature (Apusigah, 2009). The observed north–south differences in results are thus conditioned by a combination of factors related to differences in commercialization, lineage systems, gender relations, and agroecology (Dzanku et al., 2021).

With an overall lower commercialization capacity among women, it is not surprising that men's participation in output markets in terms of the gross value of sales is much higher than that of women. However, once market participation is measured in relation to total production, women's and men's commercialization rates become similar because of the low levels of output from women's plots and among FHHs.

Turning to the association between commercialization and food security, we have shown that while recent evidence might be pointing to a generally positive association between commercialization and food security, the nature of the relationship needs to be nuanced at several levels: agroecology, gender, and choice of measurement indicators. We observe no significant association between commercialization and food security in the poorer northern zone. In the highly commercialized south, however, all-year-round access to food was positively correlated with commercialization, although the magnitude of association was small, partly because most households reported access to adequate amounts of food throughout the year. Commercialization has no overall effect on diet diversity, however. As well, in the case of Asunafo North, the use of most of the arable land for cocoa meant that households were exposed to the volatility of food prices.

Across agroecological contexts, and in relation to the different food security measurements, we show that the association between commercialization and food security does not differ by sex of household head. However, at an identically high commercialization level, food tends to be more available in MHHs than FHHs, while FHHs tend to have more diverse diets. This suggests that females tend to prioritize diet quality when they control the cash from commercialization. That said, dietary diversity cannot be equated with a high-quality diet because of its ambiguity. More work is needed to improve the measurement of food security in order to strengthen the analysis of its gendered character.

Note

1 The Brong-Ahafo and Northern regions have since been split into three, with our study areas in these regions now located in the newly-created Ahafo and Savannah regions.

References

Agbosu, L., Awumbila, M., Dowuona-Hammond, C. & Tsikata, D. 2007. Customary and Statutory Land Tenure and Land Policy in Ghana, technical publication No. 70. Institute of Statistical Social and Economic Research (ISSER), University of Ghana, Accra.

Akter, S. & Basher, S. A. 2014. The impacts of food price and income shocks on household food security and economic well-being: Evidence from rural Bangladesh. *Global Environmental Change*, 25, 150–162.

Anderman, T. L., Remans, R., Wood, S. A., Derosa, K. & Defries, R. S. 2014. Synergies and tradeoffs between cash crop production and food security: a case study in rural Ghana. *Food Security*, 6, 541–554.

Apusigah, A. A. 2009. The gendered politics of farm household production and the shaping of women's livelihoods in Northern Ghana. *Feminist Africa*, 12, 51–67.

Berry, S. S. & Berry, S. 1975. *Cocoa, custom, and socio-economic change in rural Western Nigeria*, Oxford, Clarendon Press.

Boamah, F. 2014. How and why chiefs formalise land use in recent times: the politics of land dispossession through biofuels investments in Ghana. *Review of African Political Economy*, 41, 406–423.

Braimoh, A. K. 2009. Agricultural land-use change during economic reforms in Ghana. *Land Use Policy*, 26, 763–771.

Carletto, C., Corral, P. & Guelfi, A. 2017. Agricultural commercialization and nutrition revisited: empirical evidence from three African countries. *Food Policy*, 67, 106–118.

Chant, S. 2004. Dangerous equations? How female-headed households became the poorest of the poor: causes, consequences and cautions. *IDS Bulletin*, 35, 19–26.

Deininger, K. & Byerlee, D. 2011. *Rising global interest in farmland: can it yield sustainable and equitable benefits?* Washington, DC, World Bank.

Di Falco, S. & Chavas, J. P. 2009. On crop biodiversity, risk exposure, and food security in the highlands of Ethiopia. *American Journal of Agricultural Economics*, 91, 599–611.

Doss, C. R. 2002. Men's crops? Women's crops? The gender patterns of cropping in Ghana. *World Development*, 30, 1987–2000.

Doss, C. R. & Morris, M. L. 2001. How does gender affect the adoption of agricultural innovations? The case of improved maize technology in Ghana. *Agricultural Economics*, 25, 27–39.

Dzanku, F. M. & Sarpong, D. B. 2011. Agricultural diversification, food self-sufficiency and food security in Ghana: The role of infrastructure and institutions. *In:* DJURFELDT, G., ARYEETEY, E. & ISINIKA, A. C. (eds.) *African smallholders: food crops, markets and policy*, London, CAB International, pp. 189–213.

Dzanku, F. M., Tsikata, D. & Ankrah, D. A. 2021. The gender and geography of agricultural commercialisation: what implications for the food security of Ghana's smallholder farmers? *The Journal of Peasant Studies*, 48(7), 1–30.

Dzanku, F. M. & Udry, C. 2017. Flickering decades of agriculture and agricultural policy. *In:* ARYEETEY, E. & KANBUR, R. (eds.) *Ghana's economy at sixty*, Oxford, Oxford University Press.

FAO 2010. *Guidelines for measuring household and individual dietary diversity*, Rome, Food and Agriculture Organization.

FAO, IFAD & ILO 2011. *Gender dimensions of agricultural and rural employment: differentiated pathways out of poverty*, Rome, Food and Agricultural Organization.

Fischer, E. & Qaim, M. 2012. Gender, agricultural commercialization, and collective action in Kenya. *Food Security*, 4, 441–453.

Ghana Statistical Service 2018. Ghana Living Standards Survey Round 7 (GLSS 7): poverty profile in Ghana (2005–2017). Accra: Ghana Statistical Service.

Godfray, H. C. J., Beddington, J. R., Crute, I. R., Haddad, L., Lawrence, D., Muir, J. F., Pretty, J., Robinson, S., Thomas, S. M. & Toulmin, C. 2010. Food security: the challenge of feeding 9 billion people. *Science*, 327, 812–818.

Govereh, J. & Jayne, T. S. 2003. Cash cropping and food crop productivity: synergies or trade-offs? *Agricultural Economics*, 28, 39–50.

Grosz-Ngaté, M. 1997. Introduction. *In:* GROSZ-NGATÉ, M. & KOKOLE, O. (eds.) *Gendered encounters: challenging cultural boundaries and social hierarchies in Africa*, London, Routledge.

Hall, R., Scoones, I. & Tsikata, D. 2017. Plantations, outgrowers and commercial farming in Africa: agricultural commercialisation and implications for agrarian change. *The Journal of Peasant Studies*, 44, 515–537.

Hopkins, A. G. 1973. *An economic history of West Africa*, London, Longman.

Kassie, M., Ndiritu, S. W. & Stage, J. 2014. What determines gender inequality in household food security in Kenya? Application of exogenous switching treatment regression. *World Development*, 56, 153–171.

Killick, T. 1978. *Development Economics in action second edition: a study of economic policies in Ghana*, Fakenham, Norfolk, Fakenham Press Ltd.

Koizumi, T. 2015. Biofuels and food security. *Renewable and Sustainable Energy Reviews*, 52, 829–841.

Komarek, A. M. 2010. Crop diversification decisions: the case of vanilla in Uganda. *Quarterly Journal of International Agriculture*, 49, 227–242.

Lambrecht, I., Schuster, M., Samwini, S. A. & Pelleriaux, L. 2018. Changing gender roles in agriculture? Evidence from 20 years of data in Ghana. *Agricultural Economics*, 49, 691–710.

Mason, R., Ndlovu, P., Parkins, J. R. & Luckert, M. K. 2015. Determinants of food security in Tanzania: gendered dimensions of household headship and control of resources. *Agriculture and Human Values*, 32, 539–549.

Meinzen-Dick, R., Behrman, J., Menon, P. & Quisumbing, A. 2012. Gender: a key dimension linking agricultural programs to improved nutrition and health. In S. Fan and R. Pandya-Lorch, *Reshaping agriculture for nutrition and health*, Washington, DC: IFPRI, pp. 135–144.

Negash, M. & Swinnen, J. F. 2013. Biofuels and food security: micro-evidence from Ethiopia. *Energy Policy*, 61, 963–976.

Njuki, J., Kaaria, S., Chamunorwa, A. & Chiuri, W. 2011. Linking smallholder farmers to markets, gender and intra-household dynamics: Does the choice of commodity matter? *The European Journal of Development Research*, 23, 426–443.

Pierre-Louis, J. N., Sanjur, D., Nesheim, M. C., Bowman, D. D. & Mohammed, H. O. 2007. Maternal income-generating activities, child care, and child nutrition in Mali. *Food and Nutrition Bulletin*, 28, 67–75.

Pradhan, K., Dewina, R. & Minten, B. 2015. *Agricultural commercialization and diversification in Bhutan*. Washington, DC: International Food Policy Research Institute (IFPRI) and Ministry of Agriculture and Forests (MoAF).

Quisumbing, A. R., Brown, L. R., Feldstein, H. S., Haddad, L. & Peña, C. 1995. Women: the key to food security. *Food Policy Statement*, 21.

Quisumbing, A. R., Rubin, D., Manfre, C., Waithanji, E., Van Den Bold, M., Olney, D., Johnson, N. & Meinzen-Dick, R. 2015. Gender, assets, and market-oriented agriculture: learning from high-value crop and livestock projects in Africa and Asia. *Agriculture and Human Values*, 32, 705–725.

Ragasa, C., Aberman, N.-L. & Alvarez Mingote, C. 2019. Does providing agricultural and nutrition information to both men and women improve household food security? Evidence from Malawi. *Global Food Security*, 20, 45–59.

Richards, P. 1983. Farming systems and agrarian change in West Africa. *Progress in Human Geography*, 7, 1–39.

Ruel, M. T. & Alderman, H. 2013. Nutrition-sensitive interventions and programmes: how can they help to accelerate progress in improving maternal and child nutrition? *The Lancet*, 382, 536–551.

Scholes, R. J. & Biggs, R. 2004. *Ecosystem services in southern Africa: a regional assessment*, Pretoria, South Africa: Council for Scientific and Industrial Research, Millennium Ecosystem Assessment.

Shepherd, A. 1981. Agrarian change in northern Ghana: public investment, capitalist farming and famine. *In:* J. Heyer, P. Robertsand, and G. Williams (eds.) *Rural development in tropical Africa*, London, Palgrave Macmillan, pp. 168–192.

Sinclair, K., Ahmadigheidari, D., Dallmann, D. & Melgar-Quiñonez, H. 2016. Women's empowerment and food security status: a global comparative study of women living in rural and urban areas. *The FASEB Journal*, 30, 149.2–149.2.

Strasberg, P. J., Jayne, T. S., Yamano, T., Nyoro, J. K., Karanja, D. D. & Strauss, J. 1999. Effects of agricultural commercialization on food crop input use and productivity in Kenya. Michigan State University, Department of Agricultural, Food, and Resource Economics.

Timmer, C. P. 2010. Reflections on food crises past. *Food Policy*, 35, 1–11.

Torvikey, G. D., Yaro, J. A. & Teye, J. K. 2017. Farm to factory gendered employment: the case of blue skies outgrower scheme in Ghana. *Agrarian South: Journal of Political Economy*. https://doi.org/10.1177/2277976016669188

Tsikata, D. 2016. Gender, land tenure and agrarian production systems in sub-Saharan Africa. *Agrarian South: Journal of Political Economy*, 5, 1–19.

Tsikata, D. & Eweh, P. Forthcoming. *Land and agricultural commercialisation and gendered livelihoods: a synthesis of the qualitative study of four districts in Ghana*. Accra, Institute of Statistical, Social and Economic Research, ISSER Technical Publications.

Tsikata, D. & Yaro, J. A. 2014. When a good business model is not enough: land transactions and gendered livelihood prospects in rural Ghana. *Feminist Economics*, 20, 202–226.

Von Braun, J. & Kennedy, E. (eds.) 1994. *Agricultural commercialization, economic development, and nutrition*, Baltimore, MD, Johns Hopkins University Press.

Yaro, J. A., Teye, J. K. & Torvikey, G. D. 2017. Agricultural commercialisation models, agrarian dynamics and local development in Ghana. *The Journal of Peasant Studies*, 44(3), 1–17.

3 Emerging rural food markets in Kampong Thom (Cambodia)

Right to food, gender, and shifting food cultures

Fenneke Reysoo

Preamble

On our way to the villages – where we are going to conduct the interviews – the scenery of the agricultural landscape and the omnipresent entrepreneurial activities and movements of people convey an air of hope in this country that has a painful 'recent' past of genocide and civil war (Figure 3.1).

As we by-pass a man on his motorbike, transporting a bunch of fleshy, well-fed farm chickens, one of our Cambodian colleagues observes: 'The

Figure 3.1 Transport of farm chickens. © Fenneke Reysoo

DOI: 10.4324/9781003202004-5

good quality farm chickens are transported to the urban areas to be eaten by the well-to-do, and the agro-industrial skinny chickens are brought to the rural areas to be consumed by the poor'. Then, he bursts out into spontaneous laughter. The flow of chickens illustrates new ways of food exchanges. This chapter zooms in on an emerging food market, with new actors and foodstuffs, in rural Kampong Thom. It aims at stimulating a reflection on the implications of marketed food from a right to (adequate) food and gendered social justice perspective.[1] Ultimately, the readership will be able to assess by itself whether the cynical jokes and outbursts of laughter triggered by the chicken transporter can be seen as an ironical reaction to the unbearable effects of neo-liberal globalization.

Introduction

While an increasing dependency on cash and non-farm activities is a well-known pattern in agrarian change studies, in food security studies, not much attention is paid to how marketed food items are introduced in remote rural areas that transition to commercial crop cultivation. This chapter aims at providing an ethnographic account of the emergence of rural food markets in the Province of Kampong Thom. It describes a setting where rural populations and peasants increasingly have to purchase their daily food instead of producing it themselves. By building on the analytical frame of the right to food, the chapter revolves around the dimensions of availability, accessibility, and adequacy of food items that are currently sold by mobile food vendors, village grocery holders, and retailers.

The right to food refers to 'the right of everyone to have *access* to safe and nutritious food, consistent with the right to *adequate* food and the fundamental right of everyone to be free from hunger' (FAO, 2005: iii; my emphasis). Availability and accessibility refer to the means through which food is produced or acquired and the need for duty bearers to identify and eliminate structural and institutional barriers that may prevent specific groups from realizing their right to food immediately and in the future (Bourke-Martignoni, 2020, p. 151). Adequacy of food refers to 'the availability of food in a quantity and quality sufficient to satisfy the dietary needs of individuals, free from adverse substances, and acceptable within a given culture' (CESCR, 1999).

In terms of availability, this chapter presents the emergence of a variety of food outlets and looks at the range of obtainable (new) food items. The dimension of accessibility is assessed by focusing on the economic and physical constraints (money, transportation, time) of the population. The dimension of adequacy is approached through an analysis of changing food cultures, especially with regard to vernacular categories of appreciation (taste, texture, freshness, modernity). Furthermore, since issues of gender equality are directly linked to the right to food, these three dimensions of the right to food will be scrutinized by paying attention to the different effects on women and men, youth and elderly, and resource-poor and well-to-do people. Gender is understood here as a range of personal and social possibilities marked by power differentials that structure the socio-economic and political context.

Methodology

This research is the result of teamwork with researchers from various countries and disciplinary backgrounds (see Introduction). We have used a mixed-method approach by creating two data-sets. On the one hand, we collected empirical data during three qualitative interview rounds (October 2015, and January and March 2016), and on the other hand, we implemented a questionnaire-based household survey (N = 600) in 2016 and 2019 (Torrico Ramirez, 2017, 2019). For the qualitative interview rounds, we used a methodological approach inspired by the ECRIS method developed by social scientists of LASDEL in Niamey, Niger (Olivier de Sardan, 2013). The ECRIS method is very well suited for qualitative team research in remote and underserved areas (no hotel accommodation, bad roads). In a first stage, our team in Kampong Thom, composed of ten researchers/interviewers (seven men and three women), elaborated interview guides around the four topics of land commercialization, farming systems, gender relations, and food security. In a second stage, the team drove from the provincial town of Kampong Thom to the villages. Eight villages were selected based on the characteristics relevant to our overall DEMETER research project: rapid agrarian change, which – as we hypothesized – challenged gender equality (SDG5) and the right to food (related to SDG2, which is framed in terms of food security and freedom from hunger).

Except for the most remote villages where we stayed for morning and afternoon interviews, in all the other villages, we made observations and conducted the interviews in the morning (in Khmer and with audio recordings). Afterwards, we returned to our hotel, had lunch, and wrote 'hot summaries' of each interview in English. Each afternoon, our team met to discuss the insights gained and the topics to be addressed more deeply the next day (iterative process). A total of 96 semi-structured interviews were conducted with women and men, small farmers, newcomers, local traders, and authorities (Commune chiefs, councillors, and village heads). Once back in Phnom Penh, all interviews were transcribed in English and coded in the qualitative data analysis programme NVivo.[2] Eventually, patterns of food supply emerged by continuous comparisons as recommended by Glaser and Strauss (1967) in their seminal 'discovery of grounded theory'. One of these 'grounded discoveries' was the importance of mobile food vendors and village grocery stores. They are important not only for the procurement of food locally, but also for creating non-farm income-earning opportunities, especially for women. Interestingly, a food market analysis conducted in Cambodia a decade earlier by the World Food Programme found that "... only one out of five villages had a market within the village and more than 40% of the surveyed villages reported that the market is 'far away'" (WFP, 2010, p. 20). Less than a decade later, our research shows a very different picture.

Research setting

The livelihoods of the inhabitants in our research area, as in many other upland villages in Cambodia, changed dramatically over a short period. Some 20 years ago, many families were making a living from cultivating upland rice, collecting wild vegetables and plants, and hunting game from the forest. These livelihoods were profoundly affected by massive deforestation (late 1990s) and the arrival of a large-scale rubber company under a regime of economic land concessions (ELCs) granted by the Government of Cambodia in 2008 (Gironde et al., 2015; Milne and Mahanty, 2015).

In a visualization of this land transformation, using satellite images over the period 1998–2016, we can see the speed of deforestation and the expansion of ELCs in our research Communes. The reconstruction not only shows the rapid depletion of the forest and the expansion of (dirt) roads related to the ELCs, but also the speed and size of 'small-scale' deforestation over the last 20 years (Figures 3.2 and 3.3).

The migratory dynamics over the same time span are illustrated by the demographics in one of our research villages, Dang Kdar (1999–2013),

Figure 3.2 The research area in 1998
Source: Google earth, 1998/Image landsat/Copernicus

Figure 3.3 The research area in 2016
Source: Sentinelhub/Sentinel-2 L1C image on 16 April 2016

which expanded fourfold from 374 families (668 women and 614 men) in 1999 (Suon, 1999) to 1,452 families (2,873 women and 2,582 men) in 2013 (Community Database (CDB), 2014). When asked in 2016 to recall and compare the situation with five years back, almost the entire sample population reported having less access to forest areas (95% of the sample), water areas (91%), and grazing areas (89%) (Gironde et al., 2021). Strikingly, only 5.5% of the sampled households in Kampong Thom accessed the forest in the year 2018 (Torrico Ramirez, 2019; Figure 3.4).

The depletion of the forest and the closure of lands have led to at least two major changes that DEMETER team members Reysoo and Suon (2017) have captured in their article entitled 'In the forest we had plenty. Gender, Food Culture and Diet Change in Rural Cambodia (Kampong Thom Province, 2000–2015)' and Gironde et al. (2021) in their article entitled 'No cash, no food. Gendered reorganization of livelihoods and food security in Cambodia'. Both articles analyse how the rapid agrarian transition pushes smallholders out of home-food production and into commercial agriculture or wage work and how it impacts on food security of men and women, young and elderly, resource-poor and well-to-do families.

In this process of change, dietary regimes changed from the products of subsistence farming, including fishing, gathering, and hunting from the wild, to a high dependency on cash to buy food. One of the village heads, when asked to assess changes over the last 20 years, concisely remembers this transition.

> Before 2000, we were located in the logging area (Forest Concession). The logging company employed few local people, but most villagers continued to do farming inside the logging area, deep in the forest and

Figure 3.4 Deforested area and cassava fields. © Fenneke Reysoo

> mostly illegally. We needed to travel far from home, and we were afraid that the forest authority and logging company would arrest us. After 2000, we lived in the economic concession of the rubber company, and some of us could access plots of lands on an individual basis to do farming (leopard skin strategy). We then could do both farming on our land and working in the rubber company. In addition, we started to find non-farm jobs in the urban areas. More precisely, almost all families in my village have allocated some labour to non-farming activities to earn money in order to supplement the family needs – my family too. Without engaging in non-farm activities, we cannot make a good life.
>
> (Village 7, October 2015)

At the moment of the field research 2016

In 2016, a company-related infrastructure of dirt roads reached the Communes of Kraya and Beong Lvea.[3] The Commune of Kraya is composed of eight villages totalling a population of 13,644 (almost equal male: female ratio), living in, 3,189 families with an average family size of about 4.6 persons (CDB, 2014). The number of female-headed households (read women without men) varies from 15% in the village of Dang Kdar to 2% in the village of Trapeang Pring. In addition to the eight officially recognized villages, the Commune also counts six unofficial villages: new settlements, two of which are social land concessions inhabited by forcibly evicted peasants (see

Paramita, 2014). By the time of our research, six ELCs were covering 27,900 ha (ranging in size from 2,000 to 8,100 ha) (information provided by a Commune clerk, February 2016).

The Commune of Beong Lvea is composed of six official villages, with a total population of 15,807, living in 2,331 families with a variation of family size from 2.9 to 8.5 persons. In four villages, women significantly outnumber men (CDB, 2014).

The main village of Dang Kdar is situated at the crossing of two important dirt roads and has become a commercial centre with a number of micro enterprises. Another commercial centre, in the Commune of Beong Lvea, called 'Two Cottages' (Rong Pi) is situated at the bifurcation of two other important dirt roads just before entering the big rubber plantation (conceded in 2008). Four of our research villages in the Commune of Beong Lvea lay 'behind' the rubber plantation, which makes their access to 'Two Cottages' (for selling agricultural products and purchasing food) more challenging (see below). Relevant for this chapter is that

> 82% of our sampled population participates either in wage employment, self-employment or both. Accordingly, salaried and self-employment activities have become essential livelihood strategies to compensate for the reduction in cultivated areas, as well as the virtual disappearance of livestock and gathering activities.
>
> (Torrico Ramirez, 2017, p. 15)

At the provincial level, wage employment is the major source of income. The main wage labour opportunities consist of day-labour work on bigger cassava farms or on nearby rubber plantations. Other sources of income come from running a small self-employed or family business. Among the household members that participated in wage labour, 63% are male and 37% female (Torrico Ramirez, 2019, p. 34). Women's lower participation in wage labour is mainly due to their workload in reproductive activities, especially when they have to take care of children under ten years old (Gironde et al., 2021).

Market integration is thus almost universal, and people need to access food from the market because of shortfalls in their own food production. But what does the local food market look like? And what kinds of foodstuffs are made available?

Emergence of rural food markets and gender[4]

The transition to cash crops induced market activities: mobile middlemen came to the villages with trucks to buy the harvests, and some buying stations were established, especially for cassava. At the same time, other exchange activities emerged. Goods, such as fertilizers, pesticides, and weedicides; foods, such as packaged noodles and soft drinks; and services, such as beauty shops, repair shops, and karaoke bars, entered the area or were created locally to satisfy new consumer needs. In the following paragraphs, we focus on how food items are currently marketed. Interestingly, women are particularly involved in the small grocery business and mobile food vending. It is one of the self-employment niches available to women due to a number of socio-demographic, economic, and cultural determinants: lesser school accomplishments, work–family responsibilities, lack of financial start-up capital, and gender-specific cooking skills. Moreover, for mothers with (young) children, activities close to home allow them to reconcile an income-generating activity with care work. However, even if food outlets are mushrooming, their success is not self-evident and depends on financial resources for start-up investment, distance to wholesale outlets (district towns), and competitiveness. Their success is also dependent on the purchasing power of the clients. Not surprisingly, these components are highly gendered.

Mobile food vendors

In the literature, mobile food vending is mainly treated as an urban phenomenon in (big) cities of developing countries, and despite our intensive search, little to nothing is hinting at mobile food vending in rural areas. More strikingly, street food vending is almost never presented as a contribution to achieve food security and/or the right to food. Notwithstanding, a few scholars have pointed to the important function of hawkers as a source of food (Adhikari, 2011; Babb, 2018; Sohel et al., 2015; Steyn et al., 2013). In Nairobi, Sohel et al. (2015) demonstrate how important they are to offering a wide array of affordable meals to low-income households struggling with rising food prices. In this sense, vended street food can offer benefits to consumers who may lack time, money, and facilities to cook for themselves. The street food sector provides both employment for women, and inexpensive and nutritious food for the urban poor (Adhikari, 2011, p. 7). These foods 'contribute significantly' to local diets, especially among the urban poor, since they 'are convenient, cheap and easily accessible' (Steyn et al., 2013, p. 1372).

Street vendors generally have low levels of education (many of them are early school leavers) and cannot access the formal labour market due to the lack of sufficient qualifications. Many of the hawkers and food vendors in urban areas are women, and mostly 'women without men' (singles, divorcees, widows, or separated) who have to fend for their families. Thus, street vending converges at the intersection of three main domains of opportunities: no need for formal education; flexible working hours, which, in turn enables vendors to combine work and family responsibilities; and low capital requirements to start. Often, female food vendors prepare food as an extension of their usual homemaking tasks. This leads Babb (2018, p. 119 on Peru) to argue that the petty commerce of food vending, like housework, has a reproductive function in maintaining family and society.

Whereas mobile food vendors in rural areas seem to be non-existent in the literature, in our research Communes, mobile food vendors burgeon. Some,

Figure 3.5 Our research team members buying a snack from a female food vendor. © Fenneke Reysoo

both women and men, come at regular hours of the day (often early in the morning) and go from door to door. Others announce themselves through megaphones. Whereas women transport their wares on both bicycles and motorbikes, we have not observed male hawkers on bicycles, and couples sometimes use motorbikes with a side-span.[5] Mobile food vending is also an economic niche for persons with a disability (Figure 3.5).

From the villagers' perspective, the presence of mobile vendors is a welcome timesaver for those housewives who have to engage in wage work. The disadvantage, however, is that foods from the mobile vendors are not always fresh and are slightly more expensive than at the grocery stores.

Grocery stores

At the commercial centres of Dang Kdar (Kraya Commune) and 'Two Cottages' (Beong Lvea Commune), there are mechanics, beauty shops, karaoke bars, restaurants, and grocery stores. The story of one of the female grocery holders (30 years) illustrates household strategies. She started her business in 2004, because her husband did not want her to go to the field and do agricultural work when they had their first baby. A shop would allow her to combine the two tasks of raising a child and earning an income. She answered our questions while breastfeeding her youngest baby and lying in a hammock. During the interview, she continued serving customers. They purchased little things in small amounts: a motorbike driver bought petrol from a bottle, an elderly woman bought a packet of cigarettes, children bought sweets, a young woman bought a small piece of fresh cabbage, a little boy bought three packages of instant coffee, and two schoolgirls on a bike also bought sweets.

She told us that many villagers buy on credit. She lends money to customers whom she trusts and to whom she is socially connected. Generally, they pay back within ten days. Those who come from farther away and who are less well known are given fewer opportunities to buy on credit. The Pchum Ben festival (October), when families prepare big meals, is a particular moment of the year when customers ask to buy on credit.

Running a grocery store comes with entrepreneurial challenges of competition and accommodating to the seasonality of the purchasing power of the local population. Not surprisingly, stories of failure are many. A female grocery holder says that she had little money to invest and that she could not buy big stocks of merchandise at a time. In her village, others started a grocery business, and everyone sold more or less the same items, which led to market saturation (village 7). In addition, now that the road to the village exists and a more regular taxi connection to the district town of Kampong Thmar (from early morning 3 a.m.) is available, more grocery stores have been launched. The competition has become fierce and entrepreneurs with more operational investment capital, often men, bail out smaller entrepreneurs, often women.

Increased competition triggers a gender hierarchy among food vendors. The in-migrating traders are often men who have more investment capital and better connections with the outside world, as illustrated by a 34-year-old man who runs a thriving business in one of the villages. Some years back, he was a farmer with a cow cart. Sometimes, he proposed to transport firewood for others. When cutting wood in the forest became illegal, he offered other services. He invested in a van, a tractor, and a plough. This diversification led to 'time poverty' and overburdening. He chose to continue the least 'hard' work, namely wholesaling vegetables. In his van, he drives twice a week to Phnom Penh and buys many vegetables on credit (which he pays back at the end of the day by WING money transfer after his sales). The daily turnover ranges between 50 and 200 USD. His wholesale depot is located in the district town of Kampong Thmar, and his wife sells groceries at the crossroad village (with two helpers). His story illustrates the gendered dynamics of micro-entrepreneurship: men's activities tend to be more capital-based and impersonal, whereas women engage in business drawing on social relations in their nearby neighbourhoods (Reysoo, 2015; Ypeij, 2003).

Availability and accessibility of food

In rural Kampong Thom, local food markets were literally emerging under our eyes. Many interviewees hinted at the importance of food vendors for the availability of and access to food. 'Where there are people, there are food sellers' (village 8); 'Here, there is no market centre. If we need some food, we buy it from the food vendor who comes along the way from house to house' (village 5); 'When I opened the shop, there were only three or four grocery stores. Now there are many' (village 4); and 'Since the road has been connected to the village in 2014, more and more food vendors have come to the village' (village 7). These statements reflect the increasing availability of food outlets in the area.

Overall, the services of the mobile vendors are much appreciated by the buyers in the most remote and underserviced villages. Our interlocutors recall how in former days they had to travel long hours to the district market and to spend their (scarce) money on transportation. Today, with the widespread availability of bikes, motorbikes, and hand-tractors, they can more easily travel to the new commercial centres, especially the men.

Evidently, food vendors and grocery stores ensure the availability of food in the researched Communes. However, availability does not mean accessibility for all. Many of our interlocutors agree on this by using variations of the statement: 'Today you can buy anything at the market whenever you want to, provided you have money'. The latter part of the statement touches upon the issue of purchasing power and the underlying physical, structural, and institutional barriers to accessing food. Not all individuals and families have equal opportunities to buy food all year round. Indeed, access to food is becoming a steady source of stress for land-poor families who have become entirely dependent on cash to buy food. Demand for wage labourers, and hence access to cash, is seasonal and follows the cassava cultivation cycle, which has become the main cash crop in the area. An interlocutor shares, 'When people cannot sell their cassava for decent prices, they will have less to spend in the local grocery store' (village 7). Accordingly, with the volatility of cassava prices, food prices are fluctuating. The increase in the price of major food items was the most important shock experienced by 56% of the sample population. Other important shocks are the fall in the price of output (43%), increase in the price of inputs (40%) and harvest failure due to draught, flooding, or pests (38%) (Torrico Ramirez, 2019, p. 24).

The problem of food accessibility is thus closely related to the lack of money in some seasons. Related to seasonality, economically speaking, some food items are much more expensive than others, especially meat. One woman told us, 'There are petty traders who travel to my home every day. I spend a lot. Fish, pork and beef are extremely expensive'. To contextualize her statement, in the area male day labourers earn approximately 20,000–28,000 KHR (5–7 USD) per day and women 20,000–24,000 KHR (5–6 USD). Beef costs 40,000 KHR/kg (10 USD), chicken 25,000 KHR/kg (6.25 USD), pork 20,000 KHR/kg (5 USD), and fish 10,000 KHR/kg (2.5 USD). Low-quality rice costs 1,500 KHR/kg (0.375 USD) but is generally bought in bigger quantities (per bag of 50 kg). Better quality rice can go up to 3,000 KHR/kg (0.75 USD). Depending on family size, our interviewees spend around 20,000 KHR (5 USD) per day for three meals, and others report 10,000 KHR (2 USD) per day for two meals. In view of these figures, one kilogram of beef equals almost two days of wage work, and an entire day's income is spent on meals.

When money gets scarce, accessing enough food becomes difficult, and people can only buy small quantities at once. Because women bear the main responsibility for cooking, their duty of provisioning in this context comes with the anxiety of, 'Do I have enough money to cook a meal?' In addition, there is an expectation that meals must be tasty and delicious, which becomes a source of stress when money and ingredients are lacking. Time poverty also increases food insecurity. As women spend more time in productive activities away from home, both farming and non-farming, less time is available for food preparation (Johnston et al., 2015). This is deemed detrimental to family diets as highlighted by Komatsu et al. whose study shows that 'Women who spend more time cooking have more diverse diets' (2018, p. 262).

Adequacy and cultural acceptability of food

The concern with adequacy touches most importantly on the presence of the full range of food components to prepare a meal. The pillar of the three daily meals is steamed rice complemented with non-rice food (*mahoub*). Non-rice food has two elements: meat and vegetables. The entire dish gets tasty with condiments and supplements such as salt, fish sauce, soy sauce, or fermented fish (*prahok*). Freshness and taste have been and continue to be a core cultural criterion for the assessment of food quality. In earlier days, the villagers avoided eating vegetables collected the day before or meat not recently slaughtered. They were not used to storing or stockpiling food. Today, the food supplied by the mobile vendors and grocery stores is not always fresh and many complain about (potential) contamination by chemicals of the foods on offer.

The underlying question, rarely mentioned in the discussion on the adequacy of food in the realm of the right to food, is what culturally acceptable food is. The notion of cultural acceptability goes beyond the adequate quantity (as measured by food consumption scores) or quality (as measured by the household dietary diversity scores[6]) and resonates with the spirit of CESCR: 'The right to food is more than the provision of sufficient calories; it includes the cultural acceptability of food' (CESCR, 1999, *General Comment 12,* paragraph 9).

From an anthropological point of view, culturally acceptable food comes with the daily techniques of cooking and the meaning of food in a particular context. In line with Sutton (2016, pp. 349–369), who considers cooking a 'housecraft' that is mastered by women, in our research setting, it is also the women who organize labour and time to provision, plan, cook, and serve meals.[7] They have to provide the ingredients (from the land, the forests, or from the local market) and calculate the quantities. As elsewhere, in rural Cambodia, cooking methods come with particular know-how. Rice is cooked or steamed; eggs are boiled or fried; fish is fried, grilled, or dried; maize is roasted; sweet potatoes are baked, and relish is stewed (with more or less water/ simmering). When zooming in on the preparation of rice, we unveiled rules of washing the rice well and of cooking it in enough water to allow correct steaming, without draining or stirring. In the spirit of the code of conduct (*chbab srei*) that steers the decent behaviour of women, mothers hand down these skills to daughters. With age, women generally gain prestige with their cooking skills and their art to serve the meals nicely. Cultural acceptability of food comes thus with behavioural expectations for women's roles in feeding the family.

Furthermore, cultural acceptability of food comes with vernacular categories of food appreciation, which differ from scientific discourses on calories and nutrients. Scientific insights into nutrition, such as nutrients necessary for physiological development and physical activity according to gender and occupation (Golay, 2009), contrast with emic meanings of food that are organized around local categories of 'delicious' and 'not delicious' (Komatsu et al., 2018). Among the ethnic Khmer, who compose 99% of the population in our sample in Kampong Thom, a culturally acceptable meal consists of the staple rice, accompanied with relish (*mahoub*) that adds taste and facilitates swallowing (it lubricates the rice). When there is little relish, the amount of rice is increased. As long as there is rice, our interlocutors define themselves as food secure. During rice shortages, however, they have to adapt, and the last resort solution is to eat rice porridge. Referring to rice porridge and rice soup is somewhat shameful, because it reminds people of the hunger periods during the Pol Pot regime.

Our interlocutors master a semantic scheme to assess the quality of rice. They distinguish between low-quality rice (*raing chey*) and high-quality rice (*romdoul, pka malis*), the former being hard to eat, without smell, and considered to merely fill the stomach. High-quality rice is said to be 'delicious', soft, tasty, and smelling good. Hard rice ('*it is hard like stone the next day*') is cheap and eaten by the poor. It fills the stomach and gives a sense of satiation (*krob krorn).* It provides energy and power. Sar et al. (2012) relate this to the starch structure since rice starch contains 50–70% of caloric energy and is a direct source of glucose that is rapidly digested. Villagers also massively consume monosodium glutamate (MSG) as a taste enhancer. MSG is an amino-acid (code E621) and enhances taste while inhibiting hunger reactions; strikingly, 80% of the total global production of MSG is eaten in Asia (Kiéma, 2017).[8]

Overall, there is a widespread narrative among adult villagers that the 'taste is not as before'. They say that food sold in iceboxes (especially meat) is not delicious, because it lacks freshness. Some mothers also ascertained that they never buy ready-made food or cooked food (*mahoub*) from the market. If for some the argument is that they have too many children and that it would cost too much, for the majority there is the strong idea that home-cooked food tastes better. In terms of changing food cultures, we thus observe both resistance and innovation.

Indeed, meals and separate food items get new meanings. While our interlocutors continue to mobilize criteria of taste, deliciousness, freshness, and appearance (colour) to assess the adequacy of the food items, they also integrate new values, such as modernity and fanciness. In the meantime, the fear of contamination by chemicals is also on their minds, not surprisingly, since food at the local market is sourced increasingly from commercial agriculture and the food industry. Some of these food items, such as snacks, packaged food, and soft drinks, are connoted as modern and fashionable and hence desirable. These changes in food culture make me think of what Sahlins (1985) has referred to as a historical process. Thus, each time when rice is lacking or fast-food snacks are eaten, the cultural category of a 'traditional Khmer meal' is put at risk in practice, and is undergoing a transformation of its meaning. The role of children is particularly noticeable, since their desire for industrially produced fancy snacks and sweets (containing much salt, fat, and sugars) is often satisfied by adults, especially fathers, who are prone to pamper their children. What does this mean in terms of nutritious adequacy?

A closer look at the packages of some of the popular sweets illustrates the change of the nutritious food components. A 30 g package of tapioca starch/wheat flour chips contains 151 calories, of which 66 calories come from fat (90% saturated), sodium and sugar, but no vitamin A, vitamin C, calcium, or iron. It contains certified food colouring as well as antioxidants (E319, E320; E321). It is produced by LIWAYWAY Cambodia Food Industries Co Ltd. Other packages (crab crackers, green peas) of the Flying Horse Brand also contain large amounts of salt, sugar, and palm oil.[9] What we observe today in Cambodia is similar to what Richards (1939) observed in the first exhaustive anthropological study on changing food cultures (among the Mbembe, former Northern Rhodesia). Already then, she critically observed that '[I]n urban areas the new use of European foods, white flour or polished rice, have robbed the people of many of the most valuable constituents of their former diet' (p. 3).

When thinking of contemporary Cambodia, this is worrisome. In 2010, a STEP survey[10] on the prevalence of non-communicable disease risk factors was initiated on the observation that Cambodia 'is presently facing the threat of emerging obesity (…)' (2010, p. 2). People are eating below-standard quantities of fruits and vegetables, and there is a tendency to eat more meals away from home, men more than women. The nutritional transition to energy-dense nutrient-poor (EDNP) foods (Anggraini et al., 2016) occurs among populations suffering from structural or seasonal financial stress; they access EDNP foods that are more affordable and accessible than healthy foods. This explains why obesity has shifted to the poor. So on the one hand, the emerging food market brings food items to land- and resource-poor families and thus satisfies the criteria of the availability and accessibility of food as key dimensions of the right to food. On the other hand, the shift from fresh food from the forest and home-grown food to industrially processed packaged foods challenges the dimension of adequacy (in terms of nutritional components) and comes with both resistance to and acceptance of new food items.

Conclusions

> Before, farmers brought vegetables, fruit, and meat from the villages and forests to market, but now farmers bring everything from the market to villages.
>
> (village 3)

Entering the frontier area of Kampong Thom triggered a feeling of bewilderment. What had been a primary tropical forest two decades earlier has completely disappeared. Livelihoods of local people have been devastated. Local farmers have been evicted from their lands and from their home-food production. Incorporated in the market economy and commercial agriculture, many have turned into land- and money-poor masses, a kind of rural proletariat, whose access to adequate food is challenged and, therefore, their right to food susceptible to being violated.

Struck by the 'immediacy of the details' (Geertz, 1973, p. 24), we inferred that food vendors and grocery stores play a key role in feeding an increasing number of non-food farming families. 'Before' diets were more nutritious (based on home-production, collecting from the common natural resources, hunting, and fishing), and only additional condiments had to be purchased. When access to natural resources and home-production decreased, people started to buy food at the emerging local food markets, and food vendors came to the villages. With this came a number of advantages and disadvantages in terms of price, time-savings, and quality.

A range of industrially processed nutrient-poor food items were introduced, especially sweets (mostly for children), alcohol, cigarettes (mostly consumed by men), snacks, instant noodles, and soft drinks. Whereas the volume of these food supplies has increased with the growing number of grocery shops and mobile food vendors, and availability has expanded since the latter travel on a daily basis to the most remote villages, the right to food is challenged. The large majority of people do not earn enough cash to access food all year round, and the worst-off not even enough rice. In the lean season, people suffer from structural food shortages. Moreover, the adequacy of food is ambiguous. Many new food items are appreciated, not only because they are time-savers and desired, but also because of their modernity. However, there is also growing concern about the quality of purchased foods and about the chemicals that are used to produce, process, and conserve them. Middle-aged adults complain about the poor taste and lack of freshness of food from the market as compared to food formerly obtained from the forest and their own homesteads. Yet, children adore eating the sweet or salty snacks. Collaterally, their health is put at risk for non-communicable diseases, such as obesity.

Policy makers and (local) authorities do not seem to problematize this nutritional transition in terms of food accessibility and adequacy, which are two important dimensions of the right to food. They have also overlooked the important role of local food vendors, mobile or stationary, with regard to questions of food security and the right to food. Taking these activities and actors into account is important for formulating appropriate right to food strategies. Food vendors, often women, offer a wide array of affordable, accessible food items, which are increasingly the backbone for low-income households struggling with rising food and diminishing cassava prices. They also offer income-earning opportunities to women. Yet bigger, and mostly male, entrepreneurs are evicting smaller enterprises run by women, because of more investment capital, access to family labour, better means of transportation, more commercial connections, and, eventually, access to larger markets.

The main characteristic of a frontier area is that changes occur at an extraordinary rapid speed and that social and cultural dynamics evolve under our eyes. In January 2016, the Somang Reak Reay Market ('lucky and happy market') was under construction in the Commune of Beong Lvea, a huge infrastructure inviting wholesalers to the area and threatening to extinguish small local grocery holders in the villages and at the two crossroad markets – and perhaps also offering new opportunities for mobile vendors to access foodstuffs more easily and dispatch them into the remotest villages.[11]

This chapter is just a snapshot of empirical observations in the field where local peoples' livelihoods transitioned from home-produced food to commercial agriculture. The second survey conducted by our team in 2019 hints at socio-economic dynamics of peasants returning to grow their own foods. Future research is needed to confirm this trend. It appears, indeed, that, in general, female-headed households place more emphasis on the cultivation of food crops than their male counterparts do. It might be that the out-migration of men leads to a higher prevalence of female-headed households and more own food production. Expansion or reduction of local food trade may vary accordingly. But ironically, at this stage, we observe that farm-fed fleshy chickens travel from the rural areas to the cities, whereas the rural populations have largely become dependent on agro-industrially processed foods that are cheaper, thus more affordable, not always adequate, and the nutritional quality of which is impoverished.

Acknowledgements

I am grateful for the insightful comments provided by Joanna Bourke Martignoni, Christophe Gironde, Deepa Palaniappan, and Claire Somerville from the Graduate Institute, Geneva, on an earlier version of this chapter.

Notes

1 'The United Nations Millennium Development Project's Task Force on Hunger has shown that 80 per cent of the world's hungry live in rural areas' (Ziegler et al., 2011, p. 34).

2 For this chapter, we analysed the nodes 'income', 'gender relations', 'food security', and 'food culture'.

3 Until 2007, the road conditions were very bad and even worse during the rainy season (July–September). Only in 2014 did the road to access the area improve.

4 The market – in the understanding adopted here – is nothing more and nothing less than an exchange mechanism, where mutually beneficial transactions can take place. However, those without purchasing power cannot participate in the market.

5 Our survey (2019) shows that the most common means of transportation has become the motorcycle; 86% of the households in the sample reported owning at least one. Second position is held by the bicycle, in 37% of households. More men than women use motorbikes, whereas women outnumber men in using a bicycle.

6 The Household Dietary Diversity Score (HDDS) is calculated by summing equally weighted response data on the consumption of 12 different food groups in the last 24 hours. In terms of gender, it appears that men have a slightly higher dietary diversity than women in Kampong Thom (Torrico Ramirez 2019, p. 44).

7 We find that for household expenditures related to food, preparing meals, clothing and footwear, and education, the women are the main decision makers with a percentage of around 40–60%. These are the highest for food expenditures and preparing meals (55.7% and 63.96% respectively; Torrico Ramirez 2019, p. 37).

8 The American certified nutritional therapy practitioner, Craig Fear, presents a warning about the changes in food culture in Cambodia on his blog 'Fearless eating'.

> The use of MSG is {just as} rampant. Knorr is making an absolute killing over here. People use it in everything but especially soups, unfortunately. There are tons more junk foods made with the typical industrial chemicals that preserve color and flavor packaged foods. And you'll always find a section of the local market that sells this stuff; look at the food labels and you'll start to see the same ingredients over and over, namely vegetable oils, corn, wheat, sugar and a litany of chemical flavorings (such as 'natural flavors') and colorings all hidden behind bizarre, unpronounceable chemical names (azodicarbonoamide anyone?)

https://fearlesseating.net/cambodian-food/, 10.06.2020.

9 These are explicitly *halal*. *Halal* refers to an Islamic ritual to ascertain that animals are bled to death during the slaughter process. Ritually speaking, sweets cannot be *halal*. It is a pure marketing strategy.

10 STEP survey method developed by the WHO is a simple, standardized method for collecting, analysing, and disseminating data on key NCD risk factors in countries, such as trends in body-mass index, underweight, overweight, and obesity.

11 As of August 2021, the market is composed of 200–300 shops; people are selling food, meats, vegetables, clothes, and groceries. The market is led by Mr. Song Sam Ang and situated in Kampoub Ambil village.

References

Adhikari, D.B. (2011) 'Income Generation in Informal Sector: A Case Study of the Street Vendors of Kathmandu Metropolitan City', *Economic Journal of Development* Issues, Combined Issue, 13–14(1–2), pp. 1–14.

Anggraini, R., Februhartanty, J., Bardosono, S., Khusun, H. and Worsley, A. (2016) 'Food Store Choice Among Urban Slum Women Is Associated with Consumption of Energy-Dense Food', *Asia Pacific Journal of Public Health*, 28(5), pp. 458–468. Published by: Sage Publications, Inc. Stable URL: https://www.jstor.org/stable/10.2307/26686277

Babb, F.E. (2018) 'Women in the Marketplace: Petty Commerce in Peru', in Babb, F.E. (Ed.). *Women's Place in the Andes: Engaging Decolonial Feminist Anthropology*, Oakland: University of California Press, pp. 107–122.

Bourke-Martignoni, J. (2020) 'Engendering the Right to Food? International Human Rights Law, Food Security and the Rural Woman', *Transnational Legal Theory*, pp. 1–15. https://doi.org/10.1080/20414005.2018.1568789

CESCR (1999) General Comment No. 12 on the Right to Adequate Food, UN Committee of Economic, Social and Cultural Rights, UN Doc. E/C.12/1999/5.

Community Data Base Cambodia (2014) 'Raw Data Base', xlsx-file.

FAO (2005) 'Voluntary Guidelines to Support the Progressive Realization of the Right to Adequate Food in the Context of National Food Security', Adopted by the 127th Session of the FAO Council November 2004. Rome: Food and Agriculture Organization of the United Nations.

Geertz, C. (1973) *The Interpretation of Cultures. Selected Essays*, New York: Basic Books.

Gironde, C., Golay, C. and Messerli P. (ed.) (2015). *Large Scale Land Acquisitions: Focus on South-East Asia*, Leiden/Boston: Brill Nijhoff.

Gironde, C., Reysoo, F. and Torrico Ramirez, A. (2021) 'No Cash, No Food. Gendered Reorganization of Livelihoods and Food Security in Cambodia', *Journal of Peasant Studies*, online, September 5. https://doi.org/10.1080/03066150.2021.1960826

Glaser, B. and Strauss, A. (1967) *The Discovery of Grounded Theory: Strategies for Qualitative Research*, Mill Valley, CA: Sociology Press.

Golay, C. (2009) 'The Right to Food and Access to Justice: Examples at the National, Regional and International Levels', Rome: Food and Agriculture Organization of the United Nations.

Johnston, D., Stevano, S., Malapit, H., Hull, E., Kadiyala, S. (2015) Agriculture, Gendered Time Use, and Nutritional Outcomes: A Systematic Review. International Food Policy Research Institute (IFPRI), Discussion Paper 01456.

Kiéma, A. (2017). 'Les dangers du glutamate'. https://www.consoglobe.com/les-dangers-du-glutamate-cg, 10.06.2020

Komatsu, H., Malapit, H.J.L. and Theis, S. (2018) 'Does Women's Time in Domestic Work and Agriculture Affect Women's and Children's Dietary Diversity? Evidence from Bangladesh, Nepal, Cambodia, Ghana, and Mozambique', *Food Policy*, 79, pp. 256–270.

Milne, S. and Mahanty, S. (ed.) (2015) *Conservation and Development in Cambodia. Exploring Frontiers of Change in Nature, State and Society*, London/New York: Routledge.

Olivier de Sardan, J.-P. (2013) 'L'enquête socio-anthropologique de terrain. Synthèse méthodologique et recommandations à usage des étudiants', Niamey: LASDEL (Laboratoire d'études et recherches sur les dynamiques sociales et le développement local), Etudes et travaux, no. 13.

Paramita, P. (2014) 'Between Chamkar and the Kitchen: A Livelihood Approach to the Implication of Land Grabs on Food Security in Cambodian Rural Households', Geneva: Graduate Institute, MA-thesis Development Studies.

Reysoo, F. (2015) 'Inégalités de genre dans les très petites entreprises familiales: regard croisé Nord-Sud', in Droz Y. and Jacques-Jouvenot D. (Eds.). *Faire et défaire des affaires en famille*, Besançon: Presses universitaires de Franche-Comté, pp. 67–84.

Reysoo, F. and Suon, S. (2017). '"In the Forest We Had Plenty": Gender, Food Culture and Diet Change in Rural Cambodia (Kampong Thom Province, 1995–2015)', in Bock, B. and Duncan, J. (Eds.). *Gendered Food Practices from Seed to Waste*, Yearbook of Women's History 36, Hilversum, Uitgeverij Verloren, pp. 31–44.

Richards, A. (1939) *Land, Labour and Diet in Northern Rhodesia. An Economic Study of the Bemba Tribe*, London: Oxford University Press.

Sahlins, M. (1985) *Islands of History*, Chicago, IL: University of Chicago Press.

Sar, S., Gilbert, R.G. and Marks, G.C. (2012) 'Household Rice Choice and Consumption Behavior Across Agro-Climatic Zones of Cambodia', *Journal of Hunger & Environmental Nutrition*, 7(2–3), pp. 333–346. https://doi.org/10.1080/19320248.2012.707107

Sohel, A., Simiyu, E., Githiri, G., Sverdlik, A. and Mbaka, S. (2015) 'Cooking Up a Storm Community-Led Mapping and Advocacy with Food Vendors in Nairobi's Informal Settlements', London: International Institute for Environment and Development (IIED), Working Paper, 40 p.

Suon, S. (1999) 'Farming System Assessment', Kampong Thom, Provincial Department of Agriculture, Forestry and Fisheries (PDAFF).

STEPS (2010) *Prevalence of Non-Communicable Disease Risk Factors in Cambodia*, Phnom Penh: Ministry of Health.

Steyn, N.P, Mchiza, Z., Hill, J., Davids, Y.D., Venter I, Hinrichsen, E., Opperman, M., Rumbelow, J. and Jacobs, P. (2013) 'Nutritional Contribution of Street Foods to the Diet of People in Developing Countries: A Systematic Review', *Public Health Nutrition*, 17, pp. 1363–1374.

Sutton, D. (2016), 'The Anthropology of Cooking', in Klein, J.A and Watson, J.L. (Eds.). *The Handbook of Food and Anthropology*, London: Bloomsbury Academic, Chapter 16, pp. 349–369.

Torrico Ramirez, A. (2017) 'Cambodia 2016 Baseline Household Survey Report', DEMETER-R4D Research, Geneva: Graduate Institute, unpublished report.

Torrico Ramirez, A. (2019) 'Household Survey-Cambodia 2019', DEMETER-R4D Research, Geneva: Graduate Institute, unpublished report.

WFP (2010) *Cambodia Food Market Analysis and Survey Report*. Prepared by Haitham El-Noush (Independent Analyst) and World Food Programme Cambodia Country Office VAM Unit.

Ypeij, A. (2003) 'Earning More or Less: Income, Micro-Production and Women's Status in Lima's Poor Neighorhoods', in Baud, I.S.A and Post, J. (Eds.). *Realigning Actors in an Urbanizing World. Governance and Institutions from a Development Perspective*, Aldershot: Ashgate, pp. 109–118.

Ziegler, J. Golay, C., Mahon, C. and Way, S-A. (2011) *The Fight for the Right to Food: Lessons Learned*, London: Palgrave Macmillan.

4 Gender, changing food cultures, and food security in the context of agricultural commercialization in Ghana

Promise Eweh and Dzodzi Tsikata

Introduction

In Ghana, a smallholder approach to agricultural commercialization has been pursued since the colonial period (Amanor, 1999; Yaro, Teye and Torvikey, 2017). This strategy remains dominant even though Ghana was a major destination for the large-scale land acquisitions that accompanied the 2008 food, fuel, and financial crises. Over the decades, the dominance of smallholder production has been maintained through customary land tenure, and a commitment by the state to out-grower schemes and contract farming (Yaro, Teye and Torvikey, 2017). More recently, agricultural policy has favoured increased access to seeds and other inputs, as a result of a revival of input subsidy programmes that were dismantled in the 1980s (Amanor and Chichava, 2016).

Smallholder agriculture makes a significant contribution to food security. Ghana achieved the millennium development goal of reducing by half the proportion of the population living in poverty and hunger (FAO, 2015). Further, recent estimates show that only 6.4% of the Ghanaian population is undernourished (FAOSTAT, 2021). However, the multidimensional nature of food security implies that improvements in indicators such as undernourishment only offer a limited view of the entire food security situation. Even in this case where substantial progress has been made, a recent survey of four Ghanaian districts found that households faced challenges with food shortages, which were seasonal in nature and were sometimes experienced for a period of four months in a year (Dzanku et al., 2017). Further, general improvements in hunger or other dimensions of food security at the national level often overshadow substantial disparities between geographic regions and social groups. For instance, in northern Ghana where most people are not only involved in agriculture but devote most of their resources to food production, food insecurity is more severe than in the rest of the country (WFP, MOFA and GSS, 2012). Lastly, compared to men, women and children are more likely to perform poorly with respect to dimensions of food security such as food accessibility and utilization.

It is common to equate food security with zero hunger; however, it is widely accepted that food security comprises four dimensions: namely,

DOI: 10.4324/9781003202004-6

availability, accessibility, utilization, and stability of supplies. Food availability to a large extent is linked with food production since food insecurity was originally seen as a problem of inadequate food supply for an increasing human population. Thus, food insecurity was expected to be addressed through increased food production. Other aspects of food availability are 'food reserves, markets and transportation, and wild foods' (FAO et al., 2019, p. 186). However, it has been recognized since the 1980s that adequate food production at the global or national level is insufficient to meet people's food needs. Thus, it has been argued that household and individual food security is not only determined by the ability to produce own food, but crucially, by the ability to participate in market transactions. Consequently, poverty reduction and employment have been identified as critical for addressing food insecurity (Burchi and De Muro, 2016). The utilization dimension of food security draws attention to issues of malnutrition. The food security literature and policy discussions have a propensity to focus on dietary energy provided by cereals and other staples while ignoring the physiological and socio-cultural factors that influence food insecurity and malnutrition. Lastly, stability of supplies is the temporal dimension of food security that addresses the concern that 'households are food secure at all times' (FAO et al., 2019, p. 187). Food stability is influenced by a broad range of environmental, socio-economic, and political factors.

These four dimensions of food security represent the culmination of debates about the concept over the past five decades. However, a cultural dimension of food security, in addition to the four main dimensions, was introduced into the concept of food security in the 1990s (Mechlem, 2004). This recognition was reflected in the inclusion of the term 'food preferences' in the definition of food security at the World Food Summit in 1996:

> Food security, at the individual, household, national, regional and global levels is achieved when all people, at all times, have physical and economic access to sufficient, safe and nutritious food to meet their dietary needs and food preferences for an active and healthy life.
>
> (CFS, 2012, p. 5)

Similar to food security, a cultural dimension is also integral to the concept of the right to food. For example, in the report submitted to the Commission on Human Rights, the Special Rapporteur defined the right to food as

> the right to have regular, permanent and free access, either directly or by means of financial purchases, to quantitatively and qualitatively adequate and sufficient food corresponding to the cultural traditions of the people to which the consumer belongs, and which ensures a physical and mental, individual and collective, fulfilling and dignified life free of fear.
>
> (CHR, 2001, p. 7)

The recognition of culture as part of the right to food is also spelt out in several documents dealing with the concept (CESCR, 1999; FAO, 2005). Lastly, a cultural dimension of food is also recognized in the concept of food sovereignty (Bourke-Martignoni, 2020, p. 138). In spite of the acknowledgement of the cultural dimensions of food security and the right to food, research and policy addressing food cultures have been scanty. Further, little attention has been given to the interactions among food cultures, agricultural commercialization, and food security. This is compounded by the lack of a gender perspective on changing food cultures.

In this chapter, we address these gaps in the literature by applying a gender perspective to examine the relationship between changing food cultures and food security in Ghana in the context of long durée agricultural commercialization. We argue that instead of being viewed as a set of standalone elements, food cultures are intricately connected with all the four main dimensions of food security. Further, the relationships between food cultures and food security are dynamic due to different forms of commercialization occurring in different contexts of agro-ecology and social relations of class, gender, and kinship.

The chapter is primarily based on research conducted in four districts of Ghana in February 2016. Two of the districts (Asunafo North and Kwaebibirem) are in the semi-deciduous forest zone of southern Ghana, while the other two (East Gonja and Garu-Tempane) are found in the savannah zones of northern Ghana. Even though food crops such as plantain, cassava, and maize are produced in the southern districts, more productive resources were devoted to the production of non-food cash crops such as cocoa and oil palm. In the north where agro-ecological conditions do not support the production of these cash crops, the most important crops were food crops such as yam, maize, and millet. While women in the southern districts sometimes produced cash crops, their access to and control of critical resources such as land was limited, reflecting in smaller farm sizes. Similarly, men in the northern districts exercised significant control over productive resources and were recognized as the producers of food for the household. However, women were greatly involved in food production activities. Also, women in the northern districts frequently produced crops such as groundnuts, onion, and soybean, but these were regarded as less important. Thus, in both production systems, men dominated the production of crops that were assigned greater social and economic value while women were associated with crops of lower societal value.

Gender, food cultures, and food security

Food security has featured as an issue of concern in the recent literature on the implications of land and agricultural commercialization for farm households (Hall et al., 2017; Prügl et al., forthcoming). Some of this literature, which has focused on the implications for food security of different types of agricultural production arrangements, has been driven by concerns about the implications of recent large-scale land acquisitions to produce industrial cash crops and agro-fuels. Related to this are studies that have examined how changing land and labour relations in commercial agriculture are implicated in food insecurity (Yaro, Teye and Torvikey, 2017). While in some instances the gender implications of land and agricultural commercialization have been discussed, detailed research about the gendered implications of these processes has generally been lacking (Dancer and Tsikata 2015). However, gender has featured in the extensive literature on food security. This includes research that has reported better food security outcomes for male-headed households compared to their female counterparts (Kassie, Ndiritu and Stage, 2014; Tibesigwa and Visser, 2016). In other studies where these strict comparisons have been set aside for analyses of intra-household relations, findings have pointed to men's extensive control of productive resources, the exploitation of women's labour, and women's increasing work burdens as they take on more responsibilities relating to agricultural production (Kerr, 2005).

A gender analysis of food cultures contributes to discussions on food security through its focus on relationships both within and beyond the household. As a concept, food culture has received attention in various disciplines, including in the health sciences literature, which has drawn attention to the links between food cultures and health. Some of this research explores how the preference for specific foods and food preparation methods, underpinned by the social relations of class, race, gender, and kinship, are implicated in obesity and the onset of chronic non-communicable diseases such as type 2 diabetes (Williams et al., 2012).

While research from other disciplines also demonstrate possible links between food cultures and food security (Ojwang, 2011; Williams-Forson, 2014), these issues have not been taken up in the livelihoods or agrarian literature. There are a few exceptions though. For instance, Alonso, Cockx, and Swinnen (2018) have provided a focused and detailed assessment of the linkages between culture and food security based on a review of literature spanning a diverse set of case studies and disciplinary viewpoints. Based on this evaluation, the authors conclude that culture affects all the four dimensions of food security. With respect to food availability, they argue that societal conceptions about what food is, or what it is not, influence the choices of crops people produce and what they include in their diets. Culture also determines the amount and quality of food allocated to individuals at the household level. Often, this is justified on the grounds of who contributes the most to household expenditure or who performs the most difficult or important task that has benefit for the household. Again, culture's relationship with food security is argued in respect of how it mediates food utilization through specific methods of food preparation, storage, and processing at either the household or community level.

In the definitions and interpretations of food security and the right to food, issues of cultural acceptability and cooking methods among others are often highlighted as the major components of food culture. In other instances, the essence of the concept is to emphasize the important psychosocial functions performed by food other than satisfying nutritional needs (O'Kane, 2016; Williams et al., 2012). However, food cultures are also intrinsically connected with the sexual division of labour, which in many agricultural production systems determines control of, and inequalities in access to productive resources. As a result, food cultures must also be viewed as critical and dynamic channels through which gender ideologies are manifested. This is perhaps more crucial in contexts where the responsibility for food provisioning is entrusted to women. In the end, the analysis of food culture should consider its multifaceted and changing character, looking out for interactions and synergies with other dimensions of the agri-food system, such as agroecology, political economy, and the actors and relationships that underpin the system.

In this chapter, we adopt the definition of food culture that has been used within the DEMETER project. Food cultures have been defined as

> [c]onstituted by the beliefs, norms and practices about food in all its aspects – production, storage, distribution, procurement, preparation, presentation and dining arrangements and consumption. Food cultures include what is considered as food, what kinds of foods are most valued, the composition of a meal at different times of the day and year, how much protein, carbohydrates and vegetables are consumed during a meal by different members of the family or different social groups, who has responsibility for which aspects of food security and how all these influence nutrition outcomes. Food cultures are influenced by and influence geographical environment, agricultural practices, socio-economic conditions, religious beliefs, and the influence of other food cultures. Most importantly, food cultures are underpinned by social relations of class, gender, inter-generational relations, and ethnicity, and within the household, managed in the context of important social institutions such as marriage.
>
> (Tsikata, 2018)

Gendered food cultures and food security

Availability – the contours of seasonal food scarcities: incidence and determinants

Insights from feminist political ecology and political economy have highlighted how processes of land commoditization were associated with women's diminished access and control of land (Rocheleau, Thomas-Slayter and Wangari, 1996, pp. 11–12), and their inability to participate in food production (Amanor, 1999, p. 20). Similarly, findings from the southern districts of this study suggest that land and agricultural commercialization are implicated in food scarcity as more land is reserved to produce cash crops at the expense of food. At the initial stages of production, cash crops such as cocoa can be intercropped with food crops, enabling farmers to earn income from sales of such crops while directly meeting their food needs through food production. But this practice ceases when the cocoa canopies are fully formed. While there are food shortages in East Gonja, the factors shaping these shortages are different from those at play in Asunafo and Kwaebibirem. East Gonja has only one farming season, and most of its farmers produce food crops such as yam in commercial quantities. Several varieties of yam are grown and harvested at different times of the year. Participants mentioned three varieties including water yam (*Dioscorea alata L.*) and white yam (*Dioscorea rotundata Poir.*). This ensures that yam is available for cooking as food during most times of the year. Thus, the only difference between East Gonja and the southern district of Asunafo regarding food security is the lack of food diversity in the former. Our own statistical study of the four districts found that the duration of food shortages was the same for East Gonja and Asunafo (Dzanku et al., 2017).

The use of agro-chemicals was important in agricultural production in all four districts. The use of chemical fertilizers was higher in the northern districts while higher rates of pesticide and weedicide use were reported for the southern districts (Dzanku et al., 2017). However, research participants believed that agro-chemical use was resulting in increasing scarcity and extinction of vegetables and other naturally growing food. In the south, it was mentioned that the poor quality of cassava produced was due to the use of agro-chemicals. In Asunafo specifically, concerns were raised about the health and safety implications of the consumption of cabbage, which had increased in recent years. Examples of food items that were identified to be going extinct included leafy vegetables, cocoyam and its leaves (*taro*), mushrooms, bush meat in general, and snails. These were items that were formerly harvested from the commons by rural households, but they have become so scarce and expensive in local markets that they have become food for more wealthy households. A woman in Asunafo commented, 'Now the rich men are the ones who can afford to eat snails. One snail sells for GHC10, and so poor men cannot eat them'.

Food shortages were a seasonal issue and were common across all districts; however, the severity and nature of the problem varied across the districts. Consistent with other observations made earlier, the periods of shortages were different for the southern and northern districts on the one hand, and between the two northern districts on the other hand. In the south, food shortages were experienced between December/January and March whereas shortages in the north occurred between May and July (East Gonja) and May and September (Garu-Tempane). These periods of hunger occupied an important position in the social lives of people and were assigned special names. In Kwaebibirem, the period is referred to as '*ɛkɔm aba*', an Akan expression that translates literally as 'hunger has come'. In East Gonja, the period is described as '*salange*' or '*kitawunsah*',[1] which is [used] to warn households of hunger and also alert them to 'be wise in how they consume food' (Atupare, 2016). In some communities, food shortage was sometimes used to denote shortage of the most preferred food and not shortages of food in general. This was observed in the southern districts, particularly in Kwaebibirem. Thus, shortage meant shortage of, and high prices for cassava and plantain, which are used to prepare *fufu*,[2] the most important meal. In East Gonja, vegetables and 'basic ingredients such as salt, *maggi*,[3] and fish' (Atupare, 2016) were reported to be scarce during the period of shortage, although this also implied a problem with accessibility and not only that of availability. In all four districts, some individuals and social groups were more likely to experience hunger during periods of food shortages. Generally, this constituted women and children, but groups such as widows, persons with disabilities, and elderly persons lacking the support of their kin were specifically mentioned.

Many participants commented that environmental factors played an influential role in the determination of food shortages. In the south, research participants emphasized the linkage between food shortages and the weather. Accordingly, weather conditions prevented households from harvesting the crops that were used to prepare *fufu*, the most preferred staple. At the start of the major rainy season in March, strong winds pull down the plantain crop while the quality of the cassava produced has reportedly deteriorated in recent years, making it unsuitable for the preparation of *fufu*. Thus, it is reserved and used to prepare *konkonte*,[4] which is less preferred and regarded. Environmental factors were equally important in the north due to the role of the one farming season and the dry environment. In contrast, the southern districts have a major and minor rainy season, providing greater opportunities for the cultivation of crops. Beyond the physical environment are socio-cultural factors that also play an important role in food availability. For instance, in Garu-Tempane, funerals for the elderly were reported to be elaborate, sometimes taking between five and seven days to complete. During this period, food staples that are stored for household consumption are used to cater for the food needs of participants in funeral ceremonies. This was noted to lead to the depletion of food stocks and was an important cause of food scarcity during the hunger season.

Food preferences in food cultures

The most preferred foods in Kwaebibirem and Asunafo are *fufu* and *ampesi*. *Fufu* is eaten with several soups including palm nut soup, groundnut soup, and *light* or pepper soup. 'Light soup' is a broth prepared with vegetables such as garden eggs. *Fufu* is also served with soups prepared with *kontomire* (taro leaves), *ayoyo* (*Corchorus olitorius*), and *koososa* and other vegetables together with spices and protein from a variety of sources including chicken, goat meat (chevon), and smoked or salted fish. *Ampesi* is boiled yam, plantain, or cocoyam and is mostly served with garden egg (eggplant) stew or *kontomire* stew with fish as the main source of protein. Many participants considered *fufu* and *ampesi* nutritious, but these comments were mostly made in relation to the leaves and vegetables that were used to prepare the various soups. Again, *fufu* and *ampesi* were mostly consumed when the staples were available, especially in the period after harvest. During periods of food shortages, most households drastically reduce their consumption of *fufu* since cassava and plantain become scarce and expensive in the local market. Consequently, foods such as *banku*,[5] *konkonte* and yam are substituted for *fufu*. In addition to okro soup, *banku* and *konkonte* are eaten with most of the soups that are served with *fufu*. Although preparation methods and ingredients differ, *banku* and related foods such as *akple* among the Ewe and *etsew* among the Fante are important meals along the coast and other parts of Ghana. It was observed that migrants from these parts of Ghana were more likely to consider *banku* as their most preferred meal, even when cassava and plantain were abundant. This was further reflected in their reservation of more land for the growing of maize.

In the northern districts, the most preferred food was *tuo zaafi* (TZ) while yam *fufu* (yam pounded into dough) was equally important in East Gonja. TZ is prepared by cooking processed maize or millet flour (and sometimes both ingredients are mixed). It is eaten with several soups including groundnut soup and soups prepared from dried okro, *ayoyo*, *dawadawa* (*Parkia filicoides*), *neri* (*Tamarindus indica*), *kuka* (*Adansonia digitata*), and *nyankpan*. Like *fufu*, TZ is mainly a source of carbohydrates. Research participants mentioned it as the most nutritious food, especially when animal or fish protein and vegetables were part of the diet. As was observed in the south, ethnicity was quite influential in shaping the preference for food in East Gonja. For instance, the Hausa in Salaga regarded TZ as their preferred food while the Gonja in Gbung preferred yam *fufu*. Since what was considered nutritious was also shaped by preference, participants in Salaga and Gbung regarded TZ and *fufu* as the most nutritious food respectively. In East Gonja, cassava is substituted for yam during the lean season while a 'local green leafy vegetable [is] used to prepare soup' (Atupare, 2016).

In all four districts, foods that were eaten during the lean season were regarded as 'new foods' as these were not indigenous to the specific districts and were used as replacements for more preferred staples. In addition, food items such as canned tomato puree and canned fish (sardines) were also classified by some research participants as 'new foods'. However, there were instances in which the use of the term 'new foods' was restricted to food items such as polished rice and soybean kebab (specifically in the northern districts). In northern Ghana more broadly, soybean is increasingly being recognized as an important commercial crop, and the area has become more known for the promotion and increased production of the commodity than the rest of Ghana (Avornyo, Manu and Adu-Gyamfi, 2020). Soybean production is viewed as having the potential to increase youth participation in agriculture, and contract farming has been promoted as an arrangement to increase production and marketing of the commodity (Damba, Egyir and Adam, 2018). An important feature of the interest in soybean is that the effort to increase its production and consumption is driven by projects involving donors and local and international organizations (Osei-Amponsah, van Paassen and Klerkx, 2018). This new emphasis on soy, which shifts attention from local substitutes that are easier to cook to a commodity that meets the needs of agribusiness, has recruited women in their capacities as food producers and the managers of household dietary arrangements as the vectors of changing food cultures that reinforce foreign control of the agricultural value chain.

Perhaps the most notable among these 'new foods' was 'instant noodles', which was mostly consumed by children. Its consumption and the social value attached to it by adults was much lower compared with the established staples, and some participants in Kwaebibirem thought it was expensive and unhealthy (Dzanku, 2016). However, such foods may be gaining attention because they do not take long to prepare or can be purchased from local food sellers. For women, who spend a lot of time on on-farm or off-farm activities, these foods are a replacement for local dishes whose preparation require more time.

Access: responses to food shortages

The responses of households to food shortages were varied and underpinned by the availability of resources with which to purchase food, pragmatism, and a desire to survive until the season passes. In all four districts, a standard response involved adhering to one or a combination of the following measures: reducing the frequency of cooking for the household and reducing the quantity and quality of foods prepared. The last of these responses entailed reducing the amount of protein or other nutritious ingredients in diets. While these coping strategies were common in all four districts, it appears that they were applied more widely in the north, particularly in Garu-Tempane. Further, the lack of protein and nutritious ingredients in diets in Garu-Tempane was not a seasonal problem; rather, it was a critical and undesirable feature of food consumption experienced by many households throughout the year. One reason for the lack of diversity in diets was that 'most of the green vegetables and legume crops produced by the households are sold for cash income to cater for non-food needs' (Amikuzuno, 2016).

In addition to the above, other measures taken to address food shortages included the sharing and borrowing of food. In Asunafo, households borrowed food from neighbours and were sometimes required to replace it a week afterward. Similarly, food crops at local markets were sometimes purchased on credit, and payments were expected a week after such purchases. Crucially, some households requested credit from financial institutions and informal sources such as friends and clerks of cocoa-buying companies. Ensuring that food is available for the household is not the only reason why farmers borrow money, but it is quite an important one. Unfortunately, financial assistance may have negative implications for indebtedness. This was particularly noted for households that received financial support from clerks of cocoa-buying companies. These agents provide credit to cocoa farmers in return for buying their produce during the harvest season. In so doing, they take advantage of farmers' poor negotiation positions and pay them lower prices.

The practices of borrowing and sharing of food were reported in East Gonja as well, and their cultural basis seemed stronger there than in the southern districts where market transactions were much more developed. In the communities of Gbung and Grushie Zongo, oral history attests to the role of hunger in the migration and settlement of community residents at their current locations, and the importance of borrowing and sharing in ensuring the survival of both communities at difficult times in the past. Farming households in East Gonja also responded to food shortages by increasing their participation in other economic activities as they could not farm at all times. This enabled them to earn income to meet their food and other needs. The sale of livestock was, for instance, more common among men, while women focused on the production and sale of charcoal. In both northern districts, seasonal food shortages were also associated with seasonal migration, with youth labour migrating to nearby regions or the south to work for food or cash income. All in all, Kwaebibirem, despite its own share of problems, was better off than the other districts regarding food security. While some households applied some of the responses outlined above, most of the responses to food shortages in Asunafo and the northern districts were of limited significance in Kwaebibirem. Households in Kwaebibirem mostly responded to food shortages by increasing their consumption of less preferred foods such as *banku*, *konkonte*, rice, and yam. 'Participants widely indicated that they become hungry only because they are unable to eat *fufu* but that most households get adequate amounts of other foods to eat, even if not their preferred food' (Dzanku, 2016).

The purchasing and consumption of food cooked outside of the household is an important feature of food consumption in the research districts. Originally an urban phenomenon driven by convenience and perhaps cost considerations, this practice has become an important coping mechanism during periods of food shortages, when only the evening meal is cooked at home. Cooked meals and snacks are purchased for children of the house, who in any case often buy meals throughout the year while at school. Men and women who work outside the home also tend to buy cooked meals. For the women selling these cooked meals, this is an important source of income while, for consumers, buying meals frees up time for economic and social activities. Cooked meals also provide food diversity, although they sometimes raise concerns about the quality of the ingredients and the meals.

Access: the gender of food procurement practices

In many parts of rural Ghana, there is a division of responsibility in household production and procurement of food: men oversee staples or carbohydrates (linked with their ownership of the so-called family farm that demands the labour of all household members), while women are in charge of vegetables and protein, dismissively called condiments in the literature. Household members have separate purses while co-operating on household needs. In the food culture, the food is the carbohydrates and not the protein. Interestingly, discussions on food security tend to focus on staples – grains and other carbohydrates, and more rarely on vegetables and proteins. In urban Ghana, male responsibility for the staple has been converted to the institution of 'chop money', given by men to their wives either daily, weekly, or by other arrangements. What women contribute financially to food is their business, and therefore unrecognized, and yet a great source of social pressure and tension within marriage. Where families keep small animals, these are often owned by men and are used for capital expenditure rather than food. In northern Ghana, the main protein for food is small pelagic fish such as dried or smoked anchovies and herrings, unless it is a special occasion, or someone important visits. However, seasonal vegetables and legumes form an important part of the diet. In southern Ghana, there is a more varied protein source, and the 'chop money' culture fudges the division of responsibility between men and women in the procurement of food but raises the profile of men as providers.

The separation of gender roles undermines the enormous contribution made by women in food provisioning, especially in the northern districts. Generally, women are responsible for food procurement and preparation, including having to borrow or buy food on credit. In the lean season when the need for 'chop money' becomes most critical, women must sometimes take up this responsibility as men are not always able to do so. In the north, the idea that it is men who provide staples presumes that women are not involved in any farm activities. And yet, it is the produce from women's farms that are mostly reserved for use in the lean season. In East Gonja, women were assigned the more difficult role in provision of food for the household. Earlier, we noted that the food insecurity situation in East Gonja was not as severe as in Garu-Tempane due to the different varieties of yam produced and harvested at different time periods in the year. This ensured that most households had yam to cook even during the lean season. However, providing yam for the household is the responsibility of men while women are in charge of vegetables, protein, and other ingredients used to prepare soups and stews. Unfortunately, these are more difficult to come by, presenting serious difficulties for women. These have become a source of tension and domestic violence in the household (Atupare, 2016).

Utilization: gender, generation, and food-eating arrangements

In general, the preferred foods in all four districts can be seen to comprise two parts. The first part relates to the solid starch-based staple – which is considered as the main food – while the second comprises sauces or soups prepared from a variety of plants, and animal protein sources. Compared with the former, the amount of protein (especially animal protein) in the diets of most households was limited. This lack of animal protein in sauces or broths was a crucial area of difference between households in the research districts. It appeared that having smoked, dried, or salted fish in soups was a moderate improvement since poorer households sometimes relied on garden eggs, green leafy vegetables, *dawadawa*, and the other plants mentioned earlier as replacements for animal protein. However, while some households were involved in vegetable production, their main goal was to produce these crops for the market, not for household consumption. In contrast, certain households consumed more protein. For instance, a participant in one of the male focus group discussions held in Asunafo commented, 'The rich…change the type of proteins they eat. They can eat meat, fresh fish, chicken, and smoked fish alternatively. We only eat one type of protein a day. We always eat *dawadawa*, salted fish and green leafy vegetable soup' (Ankrah, 2016). In the past, protein-rich food sources such as bush meat and snail were available in the commons for collection by all people regardless of their income levels. However, because of increasing scarcity and commoditization, these have also become the prerogative of better-off households. Concerns about the nutritional value of diets were particularly pronounced in the northern districts. However, in the southern districts, the influence of social differences on food consumption was sufficiently important to be captured by several proverbs, although these sought to downplay or discount such differences. Among these proverbs is: '*Wo ti puupuu a wo bi di, ebia na eye koobi nkwan*'. This is translated as, 'If you hear someone pounding fufu, you should pound your own and eat it, because maybe they are eating theirs with just salted fish soup', meaning that you never can assume that your richer neighbours are eating higher quality food than you are. Despite a general recognition by most people that the quality of food consumed is determined by one's social position, some participants also noted that some foods eaten by wealthier households, including frozen chicken, were neither nutritious nor healthy.

Beyond the disparities between households, there are also rules about eating arrangements and food allocation and consumption that operate within the household and have important implications for growth and development, especially for children and pregnant and lactating women. Generally, men, women, and children eat separately. The best part of everything is served to the man, or men where men of the compound eat communally. This is justified on the grounds that men provide food for the household, and that they need energy to provide more food. Segregated eating arrangements and beliefs about what parts of poultry and meat to reserve for men perpetuate the privileging of male nutrition. The results from this study indicate that men continued to be privileged over women and children in food consumption. Not only were men served more food, but they were also served more protein than women and children.

Food taboos were a feature of food cultures in the research districts. Although they varied widely, they were mostly concerned with animal protein. Besides community-wide taboos that applied to all people, there were taboos that specifically exempted some individuals and groups of people from the consumption of certain foods. These proscriptions were influenced by factors such as birth order, family or clan membership, gender, and religion among others. For instance, in Garu-Tempane, it was mentioned that women (though not men) were forbidden from eating chicken or dog meat. However, with the exception of East Gonja where taboos concerning the consumption of certain foods, especially pork, were strictly adhered to probably as a result of a large Muslim population, food taboos in other districts were rarely enforced. This is in large part due to changes in society, particularly the increased role of Christianity. Thus, in general, the influence of taboos on food consumption and utilization may be limited.

Stability

Alonso, Cockx, and Swinnen (2018) note that taboos may serve the purpose of preserving specific agricultural crops or species, perhaps for their use at a later time. In Kwaebibirem, this would possibly apply to restrictions on fishing in the Apemu River, and the non-consumption of water yams among some people. However, more focused research is needed to understand the social, cultural, and economic conditions under which the food items that are tabooed at one time become culturally recognized as a legitimate source of food at another time. Another way to analyse stability within the context of food cultures is through an assessment of what are regarded as 'new' foods in various districts. In the south, although new foods such as *banku* and *konkonte* are less preferred compared to *fufu*, they ensure that households do not suffer from hunger in the lean season. In relation to this, the cultural basis of stability is embodied by the proverb '*Wo sum bodei sum kwadu*'. This was translated as 'If you plant plantain you need to plant banana as well'. Participants explained that while bananas are affected by strong winds in the same way that plantains are, they were less likely to be destroyed. This interpretation reflects thinking about how food availability and accessibility can be achieved through the production of mixed-crops and mixed-species. However, this thinking has not been reflected in the practice of devoting most of the farmlands to the production of cocoa and oil palm. As such, even though maize-based diets such as *banku* are important lean season foods, little land is devoted to its production. Commercialization is also impacting food stability through the negative effects of agro-chemicals on the safety and quality of some staples and on food bio-diversity resources. Another dimension of stability affected by commercialization is food price stability. In some communities, commercialization has also involved the giving out of land to artisanal miners. However, mining and related activities have impacted these local economies through sharp increases in food prices, which is putting small farmers and others in a vulnerable position. Further, mining was a cause of pollution of natural water resources that many vegetable farmers depend on to water their crops.

Conclusion

Food cultures represent a set of important practices and ideologies that have significant interactions with the food security dimensions of availability, accessibility, utilization, and stability of supplies. As such, food insecurity arising from land and agricultural commercialization cannot be successfully addressed unless the pivotal role of food cultures is taken into consideration in policies and interventions. In Ghana, the state, along with its donors and agribusiness are the major players influencing agricultural policy. However, their prioritization of cash crop production has been associated with seasonal food shortages and related high food prices in some communities. On the other hand, Ghana is party to several international human rights treaties that oblige it to respect, protect, and fulfil the right to food (including its cultural dimensions). Thus, efforts to meet food and nutrition needs through increased local production should occupy an important place in agricultural policy. However, interventions should target a broad range of crops rather than maintaining focus on a few select crops. As the results from this study show, food cultures are constantly changing, with households and communities shifting consumption from local staples to other foods as a coping strategy during periods of food shortages. Crop diversification does not only have the potential to stem food shortages resulting from the poor performance of particular crops but could also expand the base from which diverse nutritional elements are derived.

This essay has also examined the influence of social relations of class, gender, and generation in the experiences of food cultures and food security between and within households. Here, policy responses have only been partial, and they have had the effect of only mitigating such differences, leaving those hinging on power relations unattended. Crucial among these include structural inequalities in the distribution of productive resources between men and women. Women often experienced difficulties in gaining access to these resources, while their participation in processes central to household food provisioning was unrecognized. Women were also assigned more difficult tasks, with their inability to perform them not only being a source of stress but also domestic abuse. Commercialization has not been a reliable source of solutions to these complex challenges. Rather, it has been associated with food scarcity, the depletion of food and forest resources, and women's worsening work burdens. Opportunities for employment resulting from agribusiness operations are mainly informal while activities that were initially dominated by women and are now becoming a source of better incomes are witnessing the increasing participation of men. Meanwhile gender ideologies privilege men's contributions and nutrition to the detriment of the contributions and nutrition of women and children.

Notes

1 The term '*salange*' refers to the period of hunger between the months of June and July after farmers have completed the planting of their crops. Similarly, *kitaw-unsah* (spelled elsewhere as *kitawonsa*), is the month of July in the Akan monthly calendar. The expression means 'control their hands' since the period was characterized by food shortages (Konadu, 2012, p. 239). The use of the term in East Gonja is an indication of the level of interactions between the north and Akan settlements in the south.
2 *Fufu* is made from boiled and pounded starches (unripe plantain, cassava, cocoyam, yam). It can be made from one or two starches pounded together into a sticky dough and eaten with a variety of soups. The preparation of *fufu* is considered labour-intensive, difficult, and time-consuming by many.
3 This is a processed flavouring added to soups and stews in Ghana.
4 Fresh cassava is peeled and dried in the sun. It is then pounded and later milled into cassava flour. The flour is cooked into a solid brown dough, and served with a variety of soups.
5 *Banku* is a maize-based staple although its preparation sometimes involves the use of cassava dough as an ingredient. The process involves soaking the maize grain in water for a few days. This softens the grain for milling and also perhaps achieves a preferred texture or taste of the dough. *Banku* is distinct from *akple,* which is made from milled dried maize grain. After milling, the flour is prepared for fermentation by mixing it with water. In the end, the product of fermentation – which is a solid dough – is dissolved in water and cooked, with the addition of cassava dough.

References

Alonso, E. B., Cockx, L. and Swinnen, J. (2018) 'Culture and food security', *Global Food Security*, 17, pp. 113–127.
Amanor, K. (1999) *Global restructuring and land rights in Ghana: Forest food chains, timber, and rural livelihoods* (Research report no. 108). Uppsala: Nordic Africa Institute.
Amanor, K. S. and Chichava, S. S. (2016) 'South-South cooperation, agribusiness, and African agricultural development: Brazil and China in Ghana and Mozambique', *World Development*, 81, pp. 13–23.
Amikuzuno, J. (2016) *Qualitative research report: Garu-Tempane District.* Unpublished fieldwork report.
Ankrah, D. A. (2016) *Qualitative research report: Asunafo North District.* Unpublished fieldwork report.
Atupare, P. (2016) *Qualitative research report: East Gonja District.* Unpublished fieldwork report.
Avornyo, V. K., Manu, A. and Adu-Gyamfi, R. (2020) 'Response of soybean (Glycine max (L.) Merr) to PS-Foundation 1-0-1 biostimulant, triple super phosphate, and rhizobium inoculant', *Communications in Soil Science and Plant Analysis*, 51(7), pp. 1–12.
Bourke-Martignoni, J. (2020) 'The right to food', in Dugard, J., Porter, B., Ikawa, D. and Chenwi, L. (eds.) *Research handbook on economic, social and cultural rights as human rights.* Cheltenham: Edward Elgar Publishing, pp. 137–158.
Burchi, F. and De Muro, P. (2016) 'From food availability to nutritional capabilities: Advancing food security analysis', *Food Policy*, 60(C), pp. 10–19.

CESCR (1999) General comment No. 12: The right to adequate food (12 May 1999) E/C.12/1999/5.

CFS (2012) Coming to terms with terminology (September 2012) CFS 2012/39/4.

CHR (2001) Report by the special rapporteur on the right to food, Mr. Jean Ziegler, submitted in accordance with commission on human rights resolution 2000/10 (7 February 2001) E/CN.4/2001/53.

Damba, O. T., Egyir, I. and Adam, H. (2018) 'Drivers of farmers' relationship choice in commodity supply chains: The case of soybean in Northern Ghana', *International Journal of Irrigation and Agricultural Development*, 2(1), pp. 181–191.

Dancer, H. and Tsikata, D. (2015) *Researching land and commercial agriculture in sub-Saharan Africa with a gender perspective: Concepts, issues and methods.* Future Agricultures Consortium. Working Paper No. 132.

Dzanku, F. (2016) *Qualitative research report: Kwaebibirem District.* Unpublished fieldwork report.

Dzanku, F., Tsikata, D., Ankrah, A. D., Awo, M., Eweh, P., and Obeng, F. (2017) *Gender, livelihoods, commercialization and food security: Ghana Demeter survey report.* Unpublished fieldwork report.

FAO (2005) *Voluntary guidelines to support the progressive realization of the right to adequate food in the context of National Food Security.* Rome: FAO.

FAO (2015) *Country fact sheet on food and agriculture policy trends: Socio-economic context and role of agriculture.* Rome: FAO.

FAO, IFAD, UNICEF, WFP and WHO (2019) *The state of food security and nutrition in the world 2019: Safeguarding against economic slowdowns and downturns.* Rome: FAO.

FAOSTAT (2021) *Prevalence of undernourishment, 3-year averages.* Available from: http://www.fao.org/faostat/en/#data/FS (Accessed 16 July 2021).

Hall, R., Scoones, I. and Tsikata, D. (2017) 'Plantations, outgrowers and commercial farming in Africa: agricultural commercialisation and implications for agrarian change', *Journal of Peasant Studies,* 44(3), pp. 515–537.

Kassie, M., Ndiritu, S. W. and Stage, J. (2014) 'What determines gender inequality in household food security in Kenya? Application of exogenous switching treatment regression', *World Development*, 56, pp. 153–171.

Kerr, R.B. (2005) 'Food Security in Northern Malawi: Gender, Kinship Relations and Entitlements in Historical Context', *Journal of Southern African Studies*, 31(1), pp. 53–74.

Konadu, K. (2012) 'The calendrical factor in Akan history', *The International Journal of African Historical Studies*, 45(2), pp. 217–246.

Mechlem, K. (2004) 'Food security and the right to food in the discourse of the united nations', *European Law Journal*, 10(5), pp. 631–648.

Ojwang, D. (2011) '"Eat pig and become a beast": Food, drink and diaspora in East African Indian writing', *Research in African Literatures*, 42(3), pp. 68–87.

O'Kane, G. (2016) 'A moveable feast: Contemporary relational food cultures emerging from local food networks', *Appetite*, 105, pp. 218–231.

Osei-Amponsah, C., Paassen, A. Van and Klerkx, L. (2018) 'Diagnosing institutional logics in partnerships and how they evolve through institutional bricolage: Insights from soybean and cassava value chains in Ghana', *NJAS – Wageningen Journal of Life Sciences*, 84(1), pp. 13–26.

Prügl, E., Reysoo, F. and Tsikata, D. (2021). 'Agricultural and land commercialization – feminist and rights perspectives', *Journal of Peasant Studies,* 48(7), pp. 1419–1438.

Rocheleau, D., Thomas-Slayter, B. and Wangari, E. (1996) 'Gender and environment: A feminist political ecology perspective', in Rocheleau, D., Thomas-Slayter, B. and Wangari, E. (eds) *Feminist political ecology: Global issues and local experience*. London: Routledge, pp. 3–23.

Tibesigwa, B. and Visser, M. (2016) 'Assessing gender inequality in food security among small-holder farm households in urban and rural South Africa', *World Development*, 88(C), pp. 33–49.

Tsikata, D. (2018) *Continuities and changes in food cultures in contexts of agricultural commercialization and implications for food security outcomes: A Gender Analysis*. Presentation at a UN commission on the status of women NGO forum parallel event on agricultural commericialisation, the right to food and gender equity: Exploring the linkages, New York, 13th March, 2018.

WFP, MOFA, and GSS (2012) *Ghana comprehensive food security & vulnerability analysis 2010: Focus on Northern Ghana*. Rome: WFP.

Williams, J. D., Crockett, D., Harrison, R. L. and Thomas, K. D. (2012) 'The role of food culture and marketing activity in health disparities', *Preventive Medicine*, 55(5), pp. 382–386.

Williams-Forson, P. (2014) '"I Haven't Eaten If I Don't Have My Soup and Fufu": Cultural preservation through food and foodways among Ghanaian Migrants in the United States', *Africa Today*, 61(1), pp. 69–87.

Yaro, J. A., Teye, J. K. and Torvikey, G. D. (2017) 'Agricultural commercialisation models, agrarian dynamics and local development in Ghana', *The Journal of Peasant Studies*, 44(3), pp. 538–554.

Section II

Gender(ed) policies for food security in a commercializing world

Elisabeth Prügl

Introduction

How do local, national, and international gendered power constellations and policies influence processes of agriculture and land commercialization? The chapters in this section grapple with this question formulated in the DEMETER project. They share a political economy approach that recognizes processes of commercialization as enabled by policies and as entangled with gender and other politics in various ways.

International economic orthodoxies constitute the first layer of policies relevant for commercialization. These include colonial policies that, for example, have fostered the production of cocoa for European markets in Ghana. Today, neoliberal economic policies seek to create land markets, further industrialize agricultural production, and link small farmers into global value chains, both in Ghana and Cambodia. Gender has emerged as an issue in these policies as international agencies have insisted on the need to mainstream human rights, the World Bank, and other economic actors have framed gender equality as a matter of 'smart economics' (Prügl, 2021), and the UN has developed the notion of a right to food. These international policies find their reflections in national policies in Ghana and Cambodia in different ways.

The papers by Awo and Antwi, and by Joshi et al. explicitly interrogate how gender is being mainstreamed into policies of commercialization. There are similarities and differences that emerge when reading the papers together. Both Cambodia and Ghana seek to mainstream gender into agriculture and food security policies, and in both countries, the strategy runs into trouble. There are problems with regard to staffing, with the Cambodian bureaucracy being disproportionately male and decisions made through male networks. There are problems with funding: in Cambodia, female staff lack the seniority and access to male networks necessary to attract core funding. And there are barriers to women's meaningful participation as efforts in

DOI: 10.4324/9781003202004-7

Ghana remain mere tokenism, and civil society is fragmented. Some of this mirrors the experiences with gender mainstreaming (GM) elsewhere. But the chapters provide distinctive insights. Joshi et al. suggest that the male patronage networks that subtend the violent expropriations in Cambodia's quest for agricultural commercialization also hinder GM and the integration of gender equality considerations into agriculture and food security policies. The patriarchal political skeleton that carries commercialization thus stands in the way of efforts to achieve gender equality. Somewhat differently, Awo and Antwi suggest that the success of GM depends on opportunities for meaningful participation. Although Ghana prides itself on its democratic character, such participation is constrained for reasons ranging from fragmented civil society organizations to the routine disregard of women's representatives and interests in policy implementation. As a result, there is little attention to addressing women's needs in order to close gaping gender gaps. National policy processes thus function to marginalize women and gender equality considerations within the agenda of land and agriculture commercialization, but they do so in different ways.

Since the 1990s, development actors have promoted policies of decentralization and the governments of both Ghana and Cambodia have implemented such policies. While the intent was to democratize decision processes, decentralization has often been incomplete and has introduced a new layer of politics anchored in local power relations. In both of our case countries, there has been some administrative and participatory decentralization. Cambodia has established an elaborate sub-national planning process, drawing in sub-national state agencies. In contrast, in Ghana, policy processes seek to engage civil society, though consultations are often superficial, as confirmed in both Ghana chapters in this section. In addition, in both Cambodia and Ghana, a dearth of financing has opened the door not only to intergovernmental and government donors, but also to private companies that end up implementing commercialization policies. In Cambodia, this has empowered local patrons with connections to the party and the national elites to become conduits for international investors. In Ghana, it has empowered multinational corporations to become policy actors, as illustrated in the chapter by Adu Ankrah et al. Thus, Mondelez and Serendipalm have helped fill the funding gap, created their own policies and programmes, and co-opted government agricultural extension services for their goals. Interestingly, these multinationals have been more pro-active than the local government in implementing gender equality agendas, though their efforts have often been limited to increasing the number of women in their programmes and enterprises. Adu Ankrah et al. thus argue that they rarely address deeply gendered structures, such as land rights, and it is questionable whether their promotion of export crops at the expense of food crops is in the interest of women and their families.

In sum, the section shows that policies of commercialization and gender equality are often in tension. Efforts to mainstream gender into policy processes and enable the meaningful participation of women in such processes foundered on the shoals of patriarchal networks in Cambodia and superficial tokenism in Ghana. Financial dependencies emerged as gateways of influence for international donors, in particular at the local level, bringing with them the gospel of gender equality but also shaping a form of commercialization that responds to foreign interests.

5 Gender mainstreaming in a hybrid state

Entanglements of patriarchy and political order in Cambodia's food security sector

Saba Joshi, Elisabeth Prügl, and Muy Seo Ngouv

Introduction

Like governments around the world, Cambodia has committed to mainstreaming gender considerations into its policies, including in the area of food security. It has in place policies that recognize the role of women in fighting hunger and the connections between achieving food security and gender equality. These policies also call for an empowerment of women in agriculture and for valuing their work in the provision of food. Cambodia's GM policies thus resonate with liberal international understandings that have made gender equality a core policy goal.

Yet, Cambodia is not a liberal state. The UN-led political transition established a democratic constitution for the country in 1993, which continues to be in place. But almost 40 years of leadership of Hun Sen and the Cambodian People's Party (CPP) have created an increasingly authoritarian state. They have entrenched powerful party-based patronage networks that have served to enrich the elite (McCarthy and Un, 2017; Morgenbesser, 2018). In 2017, the CPP repressed opposition political parties, independent media and civil society, *de facto* turning the country into a one-party state. Democratic institutions, including elections held at national and local levels, thus co-exist with authoritarian practices, leading scholars to label Cambodia a hybrid state.

How does such a state advance the democratic goal of gender equality? There is little research on how illiberal governments have implemented GM, a relatively recent strategy to advance gender equality. Embraced by the UN in the 1990s, its intent is to ensure that gender becomes a consideration in all stages of policy-making and implementation in order to counteract the inadvertent reproduction of unequal gender norms. A key objective is to change institutional cultures in government that disregard the interests and needs of women and in so doing reproduce their marginalization. The strategy has received considerable critique from feminists because it often ends up inserting women into institutional hierarchies and structures without changing them. Yet, it continues to be a preferred approach to addressing gender inequality in international governance.

DOI: 10.4324/9781003202004-8

Much of the literature on GM has focused on liberal democracies and wealthy countries, or on international organizations and development interventions, where the strategy has been framed as an aspect of democratization. In these contexts, the implementation of GM has often encountered bureaucratic inertia and passive resistance. But little is known about how GM operates in politically hybrid contexts, such as contemporary Cambodia. This is particularly problematic, as authoritarian tendencies are asserting themselves globally, raising the question of how illiberal governments deal with gender equality. In this chapter, we seek to contribute such an analysis by focusing on the implementation of GM in the food security sector in Cambodia.

Food security is an area of considerable national importance in Cambodia. Around 3 million people (17.7% of the population) are considered to be poor and another 8 million are living 'near poverty' (World Bank, 2015). Approximately 2.3 million Cambodians face severe food insecurity with households spending at least 70% of their income on food, and the quality of diets remains poor (USAID, 2018, 2). Agriculture plays a key role in achieving Cambodian food security. Though the country is developing rapidly, agriculture employs at least 42% of the population (World Bank, 2018). With rice farming predominant, it provides staple food for many rural households; in addition, agricultural wage labour is increasingly becoming a source of income. Food security in Cambodia is also a gender issue. Twenty-seven percent of households are headed by women, and these are particularly likely to be poor (Maffii, 2016, ix). They have less access to resources, smaller plots of land, lower income, difficult working conditions, and lack political representation. Mainstreaming gender considerations into the food security sector thus has the potential to alleviate female poverty and strengthen women's role as farmers.

It is beyond the remit of this chapter to ascertain whether GM achieves these goals in Cambodia. Rather, we probe how the Cambodian government implements GM in the area of food security, taking into account the hybridity of its political order in interaction with hybrid gender norms. In the next section, we situate our study in the literature on GM and present our methodology. Subsequently we develop our conceptual framework by discussing Cambodia's hybrid political system and the multiple, overlapping norms related to gender equality that shape women's rights and entitlements in the field of agriculture and food security. Our empirical section explores the effects of such hybridity in institutional practices linked to GM. We first examine the way in which the entanglement of the hybrid political order with gender norms undermines efforts to correct the disproportionately male staffing of the government and slows the development of gender expertise. Second, we show how the egregious underfunding of GM is ensured through masculine patronage networks that exclude women while using their subordinate positionality to garner donor support. Finally, we discuss how civil society participation has increasingly become an empty performance as

Cambodia has transformed into a one-party state, weakening the power of feminist activists to advance gender equality through the state.

Literature review and methodology

Gender mainstreaming today consists of a dual strategy of charging sectoral administrations with integrating gender consideration into their processes, supported by strong women's machineries that oversee and monitor this integration. The literature has identified significant weaknesses in this dual structure, including a lack of resources, capacities, and expertise, but also a lack of political will, resistance, and a dearth of accountability (Hankivski, 2013; Miller and Razavi, 1998; Rai, 2003; Woodward, 2008). Typically, as the mandate moves from policy towards implementation in sectoral operations, and from national and sub-national levels, scholars have identified a process of 'evaporation' and gender concerns gradually disappearing (Kusakabe, 2005; Moser and Moser, 2005). The question of how to negotiate and counter bureaucratic resistance has become an important focus of the literature (Bustelo et al., 2016; Ferguson, 2015; Eyben and Turquet, 2013; Mukhopadhyay et al., 2007).

Research on GM is thus intensely attuned to the power politics of the strategy. Yet it often narrowly focuses on institutional and organizational processes, and the state has tended to drop out of the picture in contemporary treatments. This may be a reflection of the fact that the GM literature disproportionately focuses on countries in the EU, or countries such as Canada and Australia, with few explorations of non-Western states (Schech and Mustafa, 2010). Democracy appears thus as a background condition that apparently needs no questioning.

However, the state was a central concern of the early literature: Goetz (1997) suggested that GM bureaucracies should be conceptualized as embedded in 'political environments', and for Rai (2003), GM was about the relationship of women's organizations and the state and a matter of good governance and democratization. Indeed, the state has been an important factor in discussions of gender equality policies in Africa and Asia. Governments sometimes have embraced gender equality agendas, such as 'anti-democratic African governments' jumping on the women in development (WID) bandwagon, or 'otherwise discredited dictatorial regimes' (such as those in Ghana and Nigeria in the 1980s) pursuing 'grandiose projects' run by the wives of the heads of state, providing legitimacy to their regimes (Mama, 2007, 151). Similarly, in contemporary Northern Africa, autocratic leaders in Morocco, Tunisia, and Algeria have embraced women's rights as a way of gaining legitimacy (Tripp, 2019). A somewhat different logic seems to have motivated semi-authoritarian Malaysia when it has made the business case for gender equality, framing it as an aspect of enhancing the country's competitiveness (Elias, 2020). But in other instances, authoritarianism seems incompatible with gender equality goals. Thus, illiberal governments, such as those that have emerged recently in Europe, the United States, Brazil, or the

Philippines, are hostile to the mere reference to gender (which they frame as 'ideology'), and have attacked and dismantled gender equality machineries and legislation. Thus the character of the state matters to the way gender equality is framed and influences the kinds of policies that are advanced.

The literature on gender policies in illiberal states often explores such policies and framings. But it rarely investigates bureaucratic processes of GM that often continue to exist even in hostile policy environments. An exception is Ozkaleli (2018), who suggests that in an increasingly authoritarian state, such as Turkey, GM needs to change to become 'equity organizing' in order to account for the ethnic polarizations fostered by the regime. We hope to add additional insights through our study of Cambodia, which, like Turkey, combines liberal and illiberal characteristics. We seek to do so by focusing on the implementation of GM in one specific policy sector, that is, food security.

Our analysis draws on 15 semi-structured interviews with gender experts engaged in governmental agencies related to agriculture and food security and with civil society-based gender experts that have closely worked with or advised the government on policy-making relating to agriculture. The first 5 of these interviews were conducted from 2016 to 2017 followed by another 10 in 2020; they ranged from 60 to 90 minutes in duration.[1] Digital recordings of these interviews, obtained with participants' permission, were translated and transcribed. In some cases, the same organizations were contacted in both 2016–2017 and 2020.

Interview transcripts were coded inductively in Nvivo 12, using grounded theory methods of data analysis (see Charmaz, 2014). We coded in two rounds – first we engaged in line-by-line inductive coding, allowing the coding process to stick closely to the data. This was followed by 'focused coding', a process of refining and re-categorizing codes and drawing out themes across interviews. Memos and coding notes were developed following each round of coding, which helped capture the major reflections and ideas that emerged from the coding. To maintain the anonymity of participants, we do not state their names in this chapter.

In the following, we first draw the contours of the political order that characterizes the Cambodian state as hybrid based on secondary literature and argue that this order is entangled with a hybrid gender order. This provides us with the lens through which we then analyse our interviews, teasing out interactions of hybrid political and gender dynamics in the implementation of GM.

Political and gender orders in a hybrid state

Cambodia's political system defies easy classification; some recent labels include 'semi-authoritarian', 'semi-democratic', 'limited multi-party regime', 'civilian dictatorship', and 'party-based regime' (McCarthy and Un, 2017; Morgenbesser, 2018). Despite this diversity of classifications, several accounts concur that informal political institutions, particularly patron–client relations, play a significant role in determining political outcomes in the country.[2] Ruled by the CPP since the country's fraught transition to democracy in the early 1990s, state repression of civil society and opposition parties has accelerated since 2017. After effectively eliminating its most viable political opponents, the CPP has occupied all 125 seats in the national assembly in national elections since 2018. For many, this effectively symbolizes the 'death of democracy in Cambodia' (Sokhean, Dara, and Baliga, 2017).

Yet, while Cambodia may have entered a new era of 'hegemonic authoritarianism' (Morgenbesser, 2018), the liberal constitution of 1993 continues to endure, serving as a means to consolidate the regime's power. Thus, the judiciary practices a 'thin rule of law' to legitimize elite control (McCarthy and Un, 2017), and elections have served to reinforce and validate patron–client relationships that have historically underpinned the CPP's political authority (Morgenbesser, 2017). In other words, Cambodia's current political order represents a melange where state institutions as defined by the constitution persist but are simultaneously overlaid with networks of patron–clientelism that serve as the predominant source of political and economic power (Un, 2019).

Parallel processes have generated a hybrid gender regime that gives women strong rights to property through unwritten, customary norms on the one hand, while cementing notions of subordinate femininity in historical texts, Khmer Buddhism, and to some extent in post-civil war statutory law. Under customary norms of inheritance, Cambodian women have access to land. These, alongside other kinship practices such as uxorilocality, where the in-marrying man moves to or near his wife's natal home and typically pays or provides service to her parents, have been interpreted as markers of the 'high status' of women in Khmer society (Ledgerwood, 2018, xxii; Jacobsen, 2008, 2). According to such norms, not only do women inherit land from their parents, they also have the right to keep it in case of divorce. While both sons and daughters inherit portions of their parents' farm lands, daughters commonly inherit the family home in addition because they tend to have more care-giving obligations towards their elderly parents (Kusakabe, Yunxian, and Kelkar, 1995).

The privileging of joint marital property, codified in Cambodian law after the civil war, has somewhat weakened women's customary rights with regard to property. According to the Marriage and Family Law of 1989, all properties that the spouses gain after marriage are considered the joint property of the husband and wife, who share equal rights in owning, using, managing, and benefiting from the interests of the properties (Royal Government of Cambodia 1989, Section IV and Article 32). Women now mostly hold land jointly; while 54% of Cambodian women own land, more than two-thirds of them do so jointly (National Institute of Statistics, Directorate General for Health, and ICF International, 2014, 246–247). The practice of joint titling has accelerated this linking of land access to marital status. Introduced as a means to ensure women's equal land rights, it generates problems in the case of divorce, which often is not officially registered and may leave the names of men on women's land titles, thus limiting their rights (see Bourke Martignoni and Joshi, this volume).

Indeed, women's land rights and customary law exist in parallel to social norms that constrain women's power and rights in the family, community, and society. These are outlined in the *chbab srey* ('women's code'), a poem originating in the 19th century, which is widely memorized and reproduced in school curricula (see Anderson and Grace, 2018). It constructs the notion of the 'perfectly virtuous woman' (*srey kruap leak*), subordinate to men and obeying them irrespective of their demands or behaviour, including violence (Lilja, 2008, 70–71). Such norms often lead rural women to consider themselves inferior to men, with men entitled to more decision-making power regarding property and resources, including land ownership (Maffii, 2016, 1–4). Viewing their husbands as the heads of their families, women shoulder the burdens of taking care of children and the elderly while their active roles in farming, food production, food processing, and food distribution are overlooked and undervalued.

Liberal gender equality norms and policies, adopted by the Cambodian government in conjunction with international commitments, interact with these existing norms. The moderate rise in the percentage of women elected for office at the national (1993–2013) and sub-national levels (2002–2017), or appointed to senior government positions, has been linked to GM in government institutions and legal reforms, in addition to the advocacy efforts of civil society (Kim and Öjendal, 2014a, 26, 28). But this 'success' is not divorced from the logics of the hybrid political order. Lilja's study of female politicians in Cambodia finds that participation in electoral politics often requires family connections. Moreover, women are placed in lower ranks, in charge of stereotyped 'women's issues', and their behaviour is judged according to standards of femininity outlined in the *chbab srey* (Lilja, 2008, 62–65, 69–72). This entanglement of formal and informal gender rules also means that as politicians 'women are less connected to patronage structures and are therefore in greater need of funding to get things done' (Kim and Öjendal, 2014b, 24). Just as Cambodia's 'patriarchal state' with its 'patrimonial control over land and natural resources' shapes constructions of gender in rural areas (Beban and Bourke Martignoni, 2021), it also insinuates itself into gender equality policies.

In sum, Cambodia's hybrid political order is entangled with hybrid gender norms. The hybrid state described in the literature is a patriarchal state that encompasses a gender order with contradictory elements from religion and culture, customary and statutory laws, national and international norms. In the following, we seek to elicit how these interwoven hybriditics shape GM in the field of food security at the granular level of bureaucratic implementation.

Gender mainstreaming in Cambodian food security policy

Cambodia adopted GM as a government strategy in 1999. Under the leadership of its Ministry of Women's Affairs (MOWA), it established a gender equality plan (*Neary Rattanak*) that has been updated regularly since. 'Health and nutrition for women and girls' has been a strategic area of focus in these plans from the beginning, as has women's economic empowerment, including in rural areas. Women's nutrition also figures in the country's National Strategy for Food Security and Nutrition, in particular in objective 2, which focuses on enhancing the use and utilization of food for improving child and maternal malnutrition, eliding women's food and nutritional rights with child and maternal nutrition (CARD, 2014). In addition, the Ministry of Agriculture, Forestry and Fisheries (MAFF) has had in place a GM strategy since 2006, which was last updated in 2016. The strategy largely approaches women as farmers, and seeks to empower them economically by facilitating their 'access to goods and services for agricultural development and markets' (MAFF, 2015, 10).

MOWA oversees the mainstreaming of gender into national and sectoral policies and programmes. It is tasked to scale up GM work, support accountability mechanisms, and build capacity for the promotion of gender equality across government agencies and programmes. Most ministries have established GM action groups (GMAGs) to implement the strategy and developed associated GM action plans. MAFF's GM architecture includes a Women and Children Working Group, which is made up of representatives of the different MAFF departments.[3] The GM groups are convened regularly by MOWA.[4]

Parallel to this GM architecture is a Technical Working Group on Gender (TWG-G) which serves as a mechanism for government-donor coordination. It includes, in addition to the GMAGs, donors that work on gender, such as UNDP, UN Women, Oxfam, the Japanese International Cooperation Agency, civil society organizations, NGOs, and the private sector. The TWG-G consists of about 60 members and meets four times a year on average, co-chaired by the Minister of Women's Affairs and a representative of UNDP (Interview with Head of MOWA Gender Department, 22 May 2020).

While MOWA is the main technical agency in charge of advancing GM, there is a parallel political structure. The Cambodian National Council for Women (CNCW) was created in 2001 as an advisory body under the auspices of the Queen, and today also liaises with and reports to international bodies, such as CEDAW. It is also presided over by the Minister of Women's Affairs, and includes high-level officials from each of the sector ministries, as well as the prime minister's wife (Suon, 2015, 25f; Interview with MOWA, 17 February 2017). It thus offers a political connection between the women's machinery and the ministries. The opacity of CNCW's role in guiding GM and the personalist link it establishes between the prime minister and the state bureaucracy exemplify the blurring of political and bureaucratic authority in policy-making processes in Cambodia's hybrid state.

In the following, we demonstrate how the entangled hybridities of the Cambodian state shape the institutional practices of GM in food security. We identify two problems aggravated by the entanglement of the political order with Cambodian patriarchy: the male staffing of the bureaucracy, and the lack of resources for GM. In a third step, we discuss the limits of civil society participation in gender equality policy-making, and the failure of MOWA to serve as a potential feminist beachhead within the hybrid state.

Patriarchal staffing

The male staffing of government agencies has been a problem for the implementation of GM virtually everywhere. Because the strategy relies on the bureaucracy to apply a gender lens to its work, it needs staff that are willing to adopt such a lens. Studies have shown that this requires building capacity. Masculinist common sense, often associated with disproportionate male staffing has been widely described as an obstacle. In the literature, this problem has been diagnosed as a 'lack of political will'. In Cambodia, the lack of political will is amplified by logics of hybridity.

Male staffing is pronounced in the Cambodian government, where women made up around 41% of civil servants in 2017, but the proportion of women in decision-making positions was only 24% in 2019 (MOWA, 2019, 22). Disproportionate male staffing has been recognized as a problem, and since 2009, there has been a quota in place for 20%–50% women. Constituting a large bracket, these percentages offer considerable leeway for retaining the status quo. Moreover, the quota targets only those serving in technical positions and in jobs at the sub-national level of government. Thus, where the quota did influence the recruitment of women, they mostly have ended up in low-level positions (Interview, National Assembly, 15 May 2020). Ironically, there are no gender quotas for high-level jobs.[5]

The limited impact of gender quotas can be interpreted as an effect of patriarchal attitudes and a lack of political will. However, the reactions to MOWA's efforts to promote women into leadership positions also illustrate the impacts of the political order. MOWA's efforts encountered resistance from men, who complained that they were the victims of gender-based discrimination because they had to pay for their positions whereas the women were promoted without having to bribe their way up (Interview, MOWA, 11 February 2017). Liberal equality logics thus clashed with logics of patronage to counteract the progress of women in government. Political will emerges as structural constraint inscribed into the institutions of the hybrid state.

The effects of the continued male staffing of the state have been pernicious. While we encountered some young, male, often Western-educated staff at MOWA sincerely committed to advancing GM, men in the other ministries have shown little interest and expertise in working on gender equality. The remedy in GM typically is to create 'capacity'. However, our interviewees told us that the predominantly male staff in MAFF rarely attend technical trainings related to gender equality offered in the ministry, by MOWA or NGO partners. They considered the topic a women's issue, and their way to support it often was to support women. As one of our interviewees from MOWA recounted,

> When we talk about gender during the meetings with the representatives from ministries, they say 'I promote gender since I always assign women to participate in all gender-related meetings. They are responsible for gender'.
>
> (Interview, MOWA, 11 February 2020)

In this way, gender equality remains relegated to a different kind of politics, outside the mainstream not only of policy-making but also of masculine networks of power. Indeed, the disproportionate focus on domestic violence in Cambodian gender equality politics reinforces this understanding, precluding interventions that address structural power relations beyond the family, including those in male patronage networks that govern the allocation of land and resources (Beban and Bourke Martignoni 2021; Brickell 2020). Thus segregated, the GM bureaucracy reproduces the hybrid gender order, combining aspirations to advance equality with assuring a lack of power for women. Many of our interviewees confirmed that GM is almost exclusively a woman's job. Located in low-level positions in the line ministries, these women often lack the status and power to effect major change. The same MOWA official expressed her concern about this.

> Sometimes we worry that there will be no gender equality because there are more women involved in gender work than men. ... we will not be able to achieve gender equality if no men get involved.
>
> (Interview, MOWA, 11 February 2020)

Because, in the entangled hybrid state, power lies in the hands of men, delegating gender equality to women working at lower levels amounts to marginalizing the issue.

This refrain is, of course, familiar from other contexts, and there have been concerted efforts at the international level to win men as allies for advancing gender equality. But in Cambodia, gender specialists encounter also gender constructs that delegitimize feminist agency, and these constructs are aggravated by age, as women working in the bureaucracy tend to be younger (Interview, NGO, 11 June 2020). They run up against gender norms that expect them to act submissively.

> In our culture, it is difficult to talk about gender issues with powerful men. When we approach them, they respond saying that 'You are too young to give me advice'. We must have a strategy to talk with such persons. We can negotiate with them and be flexible .... If we start to advise them directly and tell them what to do (to support women), there is no

> way. Therefore, if we want to talk with them about gender, we should not talk directly. We need to be smart.
>
> (Interview, MOWA, 11 February 2020)

The quote illustrates the double challenge that those tasked to implement GM face. Logics of patronage make it difficult for women to reach high positions in government, and thus GM is less likely to find feminist allies in positions of power. In addition, the efforts of the gender specialists run up against constructions of submissive femininity and male superiority, which are amplified by age differences. Even if male staff were to attend them, gender training and capacity-building constitute relatively weak weapons against such norms. Thus, the male staffing of the state, the predominance of men in power, and the hybrid gender order, conspire to dull the strategy of GM, forcing gender specialists to 'be smart' and resort to subterfuge.

Patriarchal resourcing

The Cambodian state has a relatively strong planning capacity. Since the general election in 1993, in the aftermath of the UN intervention, it has undergone six planning cycles that established broad mandates and developed associated policies and instruments to monitor implementation. The Neary Rattanak national policies on gender equality follow these planning cycles, setting ambitious goals. However, the necessary budgetary allocations have not followed (Suon and Ross, 2020). This is not simply a matter of a poor state lacking resources, but a matter of a politics of resourcing that entangles equality planning with patronage and patriarchal values.

MOWA, the main agency tasked to implement the policy, has a miniscule budget of about USD 13 million, which has increased by about 5% annually over the past few years, but has remained at about 0.25% of the total national budget of Cambodia (Interview MOWA, 11 February 2020; Suon and Ross, 2020, 17). In the words of one of our interviewees at MOWA, it is a budget that went 'from nothing to less and from less to small', although she saw prospects for larger increases in the future (Interview MOWA, 11 February 2020). And while this interviewee assured us that the budget was 'ok', NGO critics disagreed: 'The big problem is money. We advocate for allocating a budget to do it, but this is not their agenda. … They're just mainstreaming … but have no budget to work on it' (Interview, NGO, 5 May 2017).

Other NGO experts pointed out that even a small budget could be sufficient for promoting gender equality, if appropriately allocated. But an interviewee argued that the MOWA channelled much of its budget towards programmes promoting women's economic empowerment at the cost of more radical feminist objectives:

> Women economic programmes do not possess a strong gender component. The way women's economic empowerment is implemented

> is patriarchal. In the execution of such projects, the roles of men and women are meant not to be equal. They do not really understand what's feminism.
>
> (Interview, NGO, 11 June 2020)

MOWA's small budget may actually be by design. In the early part of the century, MOWA's annual expenditure amounted to about 2.5% of the total national budget of Cambodia, but was dramatically reduced in 2005 from about 16 million KHR a year to 2 million, that is, to 0.35% of the national budget. Discovering the reasons for this decline proved difficult (indeed, current MOWA staff do not recall it); but some of our sources have speculated that this may have been connected to a new minister coming on board who did not have the same clout in the patronage networks. Alternatively, as one NGO activist argued, the government has used MOWA to attract donor funding, recognizing the interest of their development partners in the issue: 'MOWA has a low budget, less than the other ministries, because the government is very, very smart. They expect the MOWA to be the face of the government to deal with the donors and get money' (Interview, NGO, 1 March 2017). Indeed, Cambodia has received a massive injection of development assistance funding since 1993, including, presumably, to address gender inequality. Ironically, the promise of donor funds may have encouraged the government to underinvest in gender equality. The fact that the TWG-G is the largest of the many TWGs seeking government-donor coordination in Cambodia would support this assessment (Interview, MOWA, 11 February 2017).

While this strategy may attract international funding, it hampers the mainstreaming of gender issues into line ministries. Again, the NGO critic:

> Look at how much funding comes from UNDP, IFAD, FAO, WB, ADB to spend on these stupid things! At the same time, the Ministry of Agriculture doesn't integrate gender at all.
>
> (Interview, NGO, 1 March 2017)

Indeed, the gender work in the ministries is largely unfunded. One of our research participants at MAFF confirmed that programming and policy development work related to gender is often stalled due to lack of funding.

> There is no gender stand-alone budget line. Gender work is funded through other budget-lines such as the administrative budget. Budgeting for gender work remains a controversial issue. Thus, the Women and Children Working Group needs to shop around to implement their gender work; otherwise, the limited budget is only sufficient for organizing a small training and one event. The WCWG tries to seek support from development partners such as the Cambodian Agriculture Value Chains Program (CAVAC).
>
> (Interview, MAFF, General Directorate of Agriculture, 20 May 2020)

Funded by the Australian government, CAVAC provides a promising venue for requesting funding, but the WCWG needs to hunt for it rather than accessing it through the ministry's core budget.

Gender equality work then becomes a way to extract resources from international donors. We learned of an instance where MAFF rejected GM training because the donors would not pay per diems to those taking the training. The head of an organization working on women and development narrated:

> We were very involved in GM work for the agricultural law. We wanted to introduce training tools with the MAFF staff. The minister liked our training ideas a lot but there was a technical issue about finance so we couldn't go further. We developed this tool with an international civil society organization. To use the training tool, the Ministry of Agriculture wanted the international civil society organization to offer per diems to participants. The international organization said no per diems, only lunch. So, they couldn't do the training – it was painful.
>
> (Interview, NGO, 1 March 2017)

Rather than investing in gender equality and appreciating the offer of free capacity-building, an under-resourced bureaucracy thus used gender equality programming as a source of personal enrichment for staff.

Gender units are not the only ones to have to request funds from the general government budget and from donors, and, according to one of our MOWA interviewees (MOWA Gender Department, 22 May 2020), some are more successful than others. The fact that the GM machinery in MAFF seems to be particularly unsuccessful in securing government funding may not be unrelated to the dynamics of the patriarchal, hybrid state, in which it matters whom one knows and what networks of influence one belongs to. For the officer at MOWA, this was a matter of 'capacity', of knowing how to draw up a proposal that would be convincing to the Ministry of Economics and Finance (MEF):

> Money is there, but the next question is if you have sufficient capacity to plan or to request for it. Sometimes, the [requesting] ministry did not even draft the plan. … When they request for gifts or sarongs to be distributed during 8 March – International Women's Day, … what is the benefit?
>
> (Interview, MOWA Gender Department, 22 May 2020)

But she also recognized that it takes influence to convince one's own ministry to send a proposal to the MEF.

> The GMAG needs to lobby with [the internal] Department of Economy and Finance and their leaders… Their request may be removed [by their own ministry] … It might be due to budget constraints with certain main activities already prioritized, and the rationales the working group

> presented were not parallel or convincing. The MEF would never reject [a proposal] if the budget is there. All the working group needs is to convince their internal department to address it in the proposal.
>
> (Interview, MOWA Gender Department, 22 May 2020)

In this quote, GM or gender equality considerations are assumed not to be 'main activities' or a priority in the ministries. Instead, unfunded GMACs[6] are expected to lobby internally to have gender projects included in requests of their ministries to the MEF, so that they can get their work done. There may be a lack of capacity among GMAC members to develop such plans; but this lack of capacity must be considered in relation to the framing of the issue as marginal, the low positions of those in charge, and their lack of clout in lobbying ministerial priorities. Kept outside patriarchal networks of patronage and influence, staff charged with implementing GM struggle to insert themselves into funding processes that are intrinsically based on such networks.

This patriarchal funding structure weakens gender equality work. Thus, for example, MAFF has been hampered in developing its new GM Framework and Strategy in Agriculture. At the time of our interviews, only the General Directorate of Fisheries had finished drafting its part because it received support from the Asian Development Bank. But, as we hope to have shown in this section, the main problem hindering gender equality work is not a lack of policies; it is a lack of resourcing. One of our interviewees at MAFF confirmed that the main problem is the lack of human resources, which is directly related to a lack of core funding (Interview, General Directorate of Agriculture, 20 May 2020). And, we would argue, this lack of funding is pre-programmed in the way hierarchical male networks grease the state bureaucracy, making it difficult for the mostly female gender equality staff to get a share of the budget. In this context, international funding may contribute to further hybridizing the Cambodian gender order, but the practice of using gender equality to attract foreign funding also marginalizes the issue, excluding it from core funding streams. Patriarchal processes of resource allocation thus allow the hybrid state to perform GM for an international donor audience while undermining it from the inside.

Ambiguous allies: state–civil society interfaces in gender mainstreaming

Since 2017, with the hardening of Cambodia's one-party state, civil society organizations in the country operate on increasingly fragile grounds. In this political environment, where spaces of civic engagement have been constricted through restrictive legislation on freedom of association, surveillance, intimidation, and incarceration of civil society actors (see LICADHO, 2017, 2020), what are the possibilities for alliances and networks between feminists working in state-based agencies and those in civil society? In this section, we briefly examine the relationship between women's organizations and the hybrid state, identified in the literature as crucial for advancing gender equality agendas.

The drafting of national policies such as the National Policy on Gender Equality and the Neary Rattanak typically involve consultative meetings with various actors from the relevant ministries, NGOs, and development partners. These meetings are part of an effort of participatory policy-making that stands in contrast to informal practices of exerting influence. Our interviewees at MOWA appreciated the consultations and the contributions of NGOs to the work of governmental agencies. In the words of one, the NGOs 'had lots of good ideas', and she affirmed that these ideas were 'rarely rejected' (Interview, MOWA, 11 February 2020). Moreover, the understanding was that civil society organizations would be working on the ground, given their capacities and access:

> MOWA will work to improve capacity building in the line ministries and departments, NGOs will work at the commune level, UN Women will sponsor NGOs to work on this. This has been very helpful, and we complement each other.
>
> (Interview, MOWA Gender Department, 22 May 2020)

MOWA thus styled itself as an ally and partner of civil society with interviewees keen to establish that there were no tensions between civil society perspectives and MOWA's work on gender equality. Even with reference to points of divergence, such as the role of NGOs in highlighting the prevalence of *chbab srey* in CEDAW shadow reports, MOWA experts were quick to clarify that despite their disagreement on this issue, they did not 'blame NGOs' for raising these concerns at international fora.

> I do not care whether one works for academia, NGOs, or the government. We must work together for societal improvement and development. But let us not try to demote each other. We do not blame NGOs, but the CEDAW committee has not believed us (the government) until now.
>
> (Interview, MOWA, 11 February 2020)

Here, while the interviewee from MOWA invokes unity between the government, academia, and civil society, she suggests that the NGO shadow report to CEDAW served to 'demote' the government instead of 'working together'. This message echoes the anti-dissent and anti-critique sentiments perpetuated by the hybrid state in Cambodia where civil society is expected to perform depoliticized 'development'.

Some NGOs we spoke with voiced their dissatisfaction with the MOWA and the processes of policy consultation. Already in our 2017 interviews, an expert based in an international civil society organization in Phnom Penh questioned whether their inputs produced any impact on policy processes.

> I used to join a Technical Working Group on Gender, which was about one or two meetings per year. The discussions at these meetings are more

> like an orientation, not really seeking inputs from civil society. What they [MOWA] consider input from the civil society in policies is not clear for us. We don't see this bringing about change. Because when I talk to our [international] partners, they say that it seems like we submit inputs, but nothing changes.
>
> (Interview, NGO, 7 March 2017)

Another civil society interviewee confirmed that it was not straightforward to integrate community perspectives into policy dialogues with the MOWA. Speaking about her organization's work with women in the provinces, she recalled,

> We try to bring issues from their daily lives to the national level, to put on the agenda of the meetings of the Technical Working Group on Gender at the MOWA. Sometimes they just hear but they don't pay attention, and we have to keep raising the needs from the communities again and again.
>
> (Interview, NGO, 1 March 2017)

Thus, civil society organizations have long struggled to achieve visible impact through their advocacy efforts. Our follow-up interviews in 2020 reiterated the message that consulting with civil society was mostly a 'formality' but went further to suggest that it now also violated the democratic spirit of policy dialogues in that it simply invited civil society to rubberstamp previously made decisions.

> During the drafting process of either Neary Rattanak or NAPVAW [National Action Plan to Prevent Violence against Women], the Ministry failed to work through democratic principles as the Ministry only sought our comment after the consultants conducted their research. It seems to be just for formality's sake to prove that the Ministry has already consulted with CSOs.
>
> (Interview, NGO, 21 May 2020)

But the same interviewee also recalled instances when MOWA paid heed to their recommendations. She argued that the work of NGO networks on women's economic empowerment was integrated in the drafting of the new National Policy on Gender Equality, and MOWA invited civil society networks to lead technical working groups. In this way, civil society-based gender experts felt both validated and excluded in GM processes involving the government. Considering their ambiguous relationship, NGOs argued that their expectations for advocacy were low, but they looked forward to being pleasantly surprised when their voices were heard. 'From a civil society perspective, if we make 10 recommendations and 4 out of 10 of these are accepted, we consider this a success' (Interview, NGO, 21 May 2020).

The feminist literature shows that gender equality advocates inside and outside democratic states often form alliances to advance common goals. In the progressively more-authoritarian Cambodian state, such alliances remain weak and to an extent, perfunctory. Processes of participation maintain the outward appearance of openness and exchange, but spaces for reliable partnerships and horizontal collaboration remain narrow. In parallel, MOWA loses its function as a beachhead for advocating for gender equality; with whatever remnants of liberal democracy there are, the authoritarian state falters as an ally for realizing feminist agendas.

Conclusion

What does the case of food security in Cambodia teach us about GM in a hybrid state? And what does it illuminate about GM more broadly?

Our discussion of the Cambodian bureaucracy unveiled typical problems of disproportionately male staffing, a dearth of gender expertise, and a lack of funding. But rather than considering these a matter of a lack of political will, we hope to have shown that they are linked to institutionalized practices, what we call the hybrid political and gender orders of the Cambodian state. It is these hybridities that retain the staff of MAFF male and that of MOWA female, and that render the MOWA staff powerless; it is they that excuse male cadres from having to learn about gender equality; and it is their entangled operations that keep gender specialists outside the male patronage networks and GM off core government budgets. While the state's hybridity also has enabled alliances with civil society organizations, its authoritarian turn has easily ruptured these advocacy networks.

Considering types of state advances the understanding of how governments implement GM. It highlights how obstacles and resistances are not simply a matter of political will and attitude, but also a matter of institutionalized political and gender rules. It moreover finds these rules not only in the bureaucracy, but also beyond, in the complex common sense of a polity and its fragments of history. The Cambodian hybrid state is one such polity, and it is worth interrogating others.

Notes

1 The interviews in 2020 were conducted in Khmer by Muy Seo Ngouv and in 2016 and 2017 by Saba Joshi and Joanna Bourke Martignoni in English or Khmer (with the assistance of an interpreter). We would like to thank Joanna Bourke Martignoni for sharing her interview transcripts with us.

2 Helmke and Levitsky (2004, 727) define informal institutions as 'socially shared rules, usually unwritten, that are created, communicated, and enforced outside of officially sanctioned channels.'

Informal institutions such as patronage and clientelism are understood to be structured by socially shared rules that hinge on principles of exchange and reciprocity (Helmke and Levitsky, 2004; Scott, 1972; Stokes, 2007).

3 According to the Group's Director, only three of the members participate actively (Interview, General Directorate of Agriculture, 20 May 2020).

4 Note that the Council of Agriculture and Rural Development (CARD), which spearheads nutrition policy, is not a member of the GMAG group assembled by

MOWA. This emerged in our interviews with MOWA and prompted a promise to rectify the matter.

5 There also are no gender quotas for the legislature. After the dissolution of the main opposition party in 2017, women held only 15% of seats in the Cambodian parliament. Leonie, Kijewski (November 2017). *Only Two Women Join the National Assembly.* The Phnom Penh Post. Retrieved from https://www.phnompenhpost.com/national-politics/only-two-women-join-national-assembly.

6 Whose members 'mostly are occupied by their core tasks, the tasks on gender is just an accessory' (Interview, MOWA Gender Department, 22 May 2020).

References

Anderson, E. and Grace, K. (2018). 'From Schoolgirls to "Virtuous" Khmer Women: Interrogating Chbab Srey and Gender in Cambodian Education Policy.' *Studies in Social Justice*, 12 (2), 215–234. https://doi:10.26522/ssj.v12i2.1626.

Beban, A. and Bourke Martignoni, J. (2021). '"Now the Forest Is Over": Transforming the Commons and Remaking Gender in Cambodia's Uplands.' *Frontiers in Sustainable Food Systems*, 5. https://www.frontiersin.org/article/10.3389/fsufs.2021.700990.

Brickell, K. (2020). *Home SOS: Gender, Violence, and Survival in Crisis Ordinary Cambodia.* Hoboken, NJ and Chichester, West Sussex: John Wiley & Sons.

Bustelo, M., Ferguson, L. and Forest, M. (eds.) (2016). *The Politics of Feminist Knowledge Transfer: Gender Training and Gender Expertise.* Houndmills, Basingstoke, Hampshire: Palgrave Macmillan.

Charmaz, K. (2014). *Constructing Grounded Theory.* 2nd edition. Introducing Qualitative Methods. London: Thousand Oaks, CA: Sage.

Council for Agricultural and Rural Development (CARD) (2014). 'National Strategy for Food Security and Nutrition (NSFN) 2014–2018.' Phnom Penh, Cambodia: Council for Agricultural and Rural Development.

Elias, J. (2020). *Gender Politics and the Pursuit of Competitiveness in Malaysia: Women on Board.* https://public.ebookcentral.proquest.com/choice/publicfullrecord.aspx?p=6132266.

Eyben, R. and Turquet, L. (eds.) (2013). *Feminists in Development Organizations: Change from the Margins.* Bourton on Dunsmore: Practical Action.

Ferguson, L. (2015). 'This Is Our Gender Person.' *International Feminist Journal of Politics*, 17 (3): 380–397. https://doi.org/10.1080/14616742.2014.918787.

Goetz, A.M. (1997). *Getting Institutions Right for Women in Development.* London: Zed Books.

Hankivski, O. (2013). 'Gender Mainstreaming: A Five-Country Examination.' *Politics & Policy*, 41 (5): 629–655. https://doi.org/10.1111/polp.12037.

Helmke, G. and Levitsky, S. (2004). 'Informal Institutions and Comparative Politics: A Research Agenda.' *Perspectives on Politics*, 2 (4): 725–740. https://doi.org/10.1017/S1537592704040472.

Jacobsen, T. (2008). *Lost Goddesses: The Denial of Female Power in Cambodian History.* Copenhagen: NIAS Press.

Kim, S. and Öjendal, J. (2014a). 'Gendering Decentralization in Cambodia.' Research Report 4. Research Report. Malmo: Swedish International Centre for Local Democracy (ICLD).

———. (2014b). 'Gatekeepers in Local Politics: Political Parties in Cambodia and Their Gender Policy.' Working Paper No. 87. Working Paper Series. Phnom Penh, Cambodia: Cambodia Development Resource Institute (CDRI).

Kusakabe, K. (2005). 'Gender Mainstreaming in Government Offices in Thailand, Cambodia, and Laos: Perspectives from Below.' *Gender and Development*, 13 (2): 46–56.

Kusakabe, K., Yunxian, W. and Kelkar, G. (1995). 'Women and Land Rights in Cambodia.' *Economic and Political Weekly*, 30 (43): WS87–WS92.

Ledgerwood, J. (2018). 'Introduction.' In Ebihara, M.M. (ed.), *Svay: A Khmer Village in Cambodia*. Edited by A. Mertha. Ithaca, NY: Cornell University Press.

LICADHO (2017). 'The Dangers of Dissent: Attacks on Human Rights Defenders'. Briefing Paper, Phnom Penh: LICADHO. https://www.licadho-cambodia.org/reports/files/2232017_HRD_ENG[finalWeb].pdf

——— (2020). 'The Fight for Freedom: Attacks on Human Rights Defenders 2018–2020', Briefing Paper, Phnom Penh: LICADHO. https://www.licadho-cambodia.org/reports/files/23320201218_Human%20Rights%20Defenders%20Report%20 2018_2020_EN.pdf

Lilja, M. (2008). *Power, Resistance and Women Politicians in Cambodia: Discourses of Emancipation*. Copenhagen, Abingdon: NIAS.

MAFF (2015). *Gender Mainstreaming Policy and Strategic Framework in Agriculture 2016–2020*. Phnom Penh: Ministry of Agriculture, Forestry and Fishery.

Maffii, M. (2016). *Summary of the Impact of Gender Policy Processes on the Right to Food: The Case of Cambodia*. Rome: FAO. http://www.fao.org/3/a-i5489e.pdf.

Mama, A. (2007). 'Critical Connections: Feminist Studies in African Contexts.' In Cornwall, A., Harrison, E., and Whitehead, A. (eds.), *Feminisms in Development: Contradictions, Contestations and Challenges*. London: Zed Books, pp. 150–160.

McCarthy, S. and Un, K. (2017). 'The Evolution of Rule of Law in Cambodia.' *Democratization*, 24 (1): 100–118. https://doi.org/10.1080/13510347.2015.1103736.

Miller, C. and Razavi, S. (1998). *Missionaries and Mandarins: Feminist Engagement with Development Institutions*. Intermediate Technology Publications in association with the United Nations Research Institute for Social Development.

Morgenbesser, L. (2017) 'The Failure of Democratisation by Elections in Cambodia.' *Contemporary Politics*, 23 (2): 135–155. https://doi.org/10.1080/13569775.2016.1230317.

———. (2018). 'Misclassification on the Mekong: The Origins of Hun Sen's Personalist Dictatorship.' *Democratization*, 25 (2): 191–208. https://doi.org/10.1080/13510347.2017.1289178.

Moser, C.O.N. and Moser, A. (2005). 'Gender Mainstreaming Since Beijing: A Review of Success and Limitations in International Institutions.' *Gender & Development*, 13 (2): 11–22. https://doi.org/10.1080/13552070512331332283.

MOWA (2019) *Cambodia Report: On the Occasion of the 25th Anniversary of the Fourth World Conference on Women and the Adoption of the Beijing Declaration and Platform for Action (1995)*. Phnom Penh, Cambodia: Ministry of Women's Affairs.

Mukhopadhyay, M. and Wong, F. (2007). *Revisiting Gender Training: The Making and Remaking of Gender Knowledge: A Global Sourcebook*. Amsterdam, The Netherlands; Oxford: KIT Publishers; Oxfam Publishing.

National Institute of Statistics, Directorate General for Health, and ICF International (2014). 'Cambodia Demographic and Health Survey.' Phnom Penh, Cambodia and Rockville, Maryland, USA: National Institute of Statistics, Directorate General for Health, and ICF International.

Ozkaleli, U. (2018). 'Intersectionality in Gender Mainstreaming: Equity Organizing in Turkey.' *Journal of Women, Politics & Policy*, 39 (2) (April): 127–150. https://doi.org/10.1080/1554477X.2018.1449524.

Prügl, E. (2021). 'Smart Economics.' In Berik, G., and Kongar, E. (eds.), *The Routledge Handbook of Feminist Economics*. London and New York: Routledge, pp. 468–476.

Rai, S. (2003). *Mainstreaming Gender, Democratizing the State? Institutional Mechanisms for the Advancement of Women*. Manchester: Manchester University Press.

Schech, S. and Mustafa, M. (2010). 'The Politics of Gender Mainstreaming Poverty Reduction: An Indonesian Case Study.' *Social Politics: International Studies in Gender, State & Society*, 17 (1) (March): 111–135. https://doi.org/10.1093/sp/jxp025.

Scott, J.C. (1972). 'Patron-Client Politics and Political Change in Southeast Asia.' *The American Political Science Review*, 66 (1): 91. https://doi.org/10.2307/1959280.

Sokhean, B., Dara, M. and Baliga, A. (2017 November). '"Death of Democracy": CNRP Dissolved by Supreme Court Ruling | Phnom Penh Post.' *Phnom Penh Post*, 17, Online edition. https://www.phnompenhpost.com/national-post-depth-politics/death-democracy-cnrp-dissolved-supreme-court-ruling.

Stokes, S.C. (2007). 'Political Clientelism.' In Stokes, S.C. and Boix, C. (eds.), *The Oxford Handbook of Comparative Politics*. Oxford: Oxford University Press, pp. 604–627.

Suon, S. (2015). Review of Responsive Laws, Policies and Strategies toward Gender Equality in Cambodia. Unpublished draft.

Suon, S. and Ross, S. (2020). 'A Review of Cambodian Law, Policies, and Politics Related to Land Governance, Gender Equality and Food Security.' DEMETER Working Paper 1/2020.

Tripp, A.M. (2019). *Seeking Legitimacy: Why Arab Autocracies Adopt Women's Rights*. Cambridge: Cambridge University Press.

Un, K. (2019). *Cambodia: Return to Authoritarianism*. 1st edition. Elements in Politics and Society in Southeast Asia. Cambridge University Press. https://doi.org/10.1017/9781108558648.

USAID (2018). 'Cambodia: Nutrition Profile.' USAID. 24/3/2021. https://www.usaid.gov/sites/default/files/documents/1864/Cambodia-Nutrition-Profile-Mar2018-508.pdf

Woodward, A.E. (2008). 'Too Late for Gender Mainstreaming? Taking Stock in Brussels.' *Journal of European Social Policy*, 18 (3): 289–302. https://doi.org/10.1177/0958928708091061.

World Bank (2015). 'Cambodian Agriculture in Transition: Opportunities and Risks.' Economic and Sector Work Report No. 96308-KH. Agriculture and Global Practice, East Asia and Pacific Region. Washington, DC: The World Bank. http://documents.worldbank.org/curated/en/805091467993504209/pdf/-96308-ESW-KH-White-cover-P145838-PUBLIC-Cambodian-Agriculture-in-Transition.pdf

———. (2018). *Cambodia: Overview*. Retrieved on 14 May 2018, from http://www.worldbank.org/en/country/cambodia/overview.

6 Minding the gap in agriculture and food security

Gender mainstreaming and women's participation in policy processes in Ghana

Martha A. Awo and Anna Antwi

Introduction

After years of criticism for ignoring gender concerns in development, there has been a paradigm shift towards making policies more inclusive. Key among the wave of gender equality policies of the 1970s, 1980s, and 1990s is the 1995 Beijing Platform for Action that introduced gender mainstreaming and emphasized the need to restructure policies and integrate gender into development processes (UN, 1979). The intention was to create awareness of gender biases and injustices in structures and institutionalized cultures and to remove them (Woodford-Berger, 2007). The strategy has been taken up by governments, including the mainstreaming of gender into agricultural production and food security policies. The Comprehensive Africa Agriculture Development Programme (CAADP) and the Alliance for a Green Revolution in Africa (AGRA), for example, have gender equity components.

Despite such efforts, there continue to be global gender gaps in agriculture with regard to vulnerabilities, access to resources, and productivity (FAO, 2011 and Huyer, 2016). Many development programmes and policies continue to be designed in a way that uses male farmers as the norm (World Bank et al., 2009). This is notwithstanding evidence from a body of research confirming the contributions women make to agriculture and food security (FAO, 2011 and Huyer, 2016). Thus, studies show that, in developing countries, women constitute approximately 43% of the agricultural labour force, and the FAO's State of Food and Agriculture report reveals that closing the gender gap in agriculture could lift 150 million people out of hunger (FAO, 2011 and Huyer, 2016). According to a study led by the World Bank, the estimated cost of the gender gap in agricultural productivity is $100 million in Malawi, $105 million in Tanzania, and $67 million in Uganda (UN Women, UNDP, UNEP, & World Bank, 2015).

In Ghana, about 52% of the female population are engaged in agriculture, with women accounting for close to 70% of total food production (JICA, 2013). Moreover, 46.4% of women own or operate a farm (GSS, 2019, p. 162).

DOI: 10.4324/9781003202004-9

However, their contribution is hardly recognized (Tsikata, 2016). Thus, Apusigah finds, from a study in Northern Ghana, that women are seen as farm hands who play specialized roles in on-farm activities, and non-farm hands are perceived largely as non-productive and at best as reproductive (Apusigah, 2009). This leads to gaps in women's access to and control over productive resources. Gender analysis conducted in the agriculture sector has identified gender gaps in institutional arrangements and implementation procedures; in access to and control over land; and in access to credit and financial services (NETRIGHT, 2014). Land has been a long-standing issue of debate aggravated by a lack of gender-disaggregated data required to inform policy. The report on the seventh round of the Ghana Living Standard Survey (GSS, 2019), for instance, does not disaggregate modes of land acquisition by sharecropping, inheritance, tenancy, purchase, right of user or husband's land. In terms of access to credit, the report shows that only 6.2% of the female population, as compared to 22.6% of men, are likely to get access to agricultural credit (GSS, 2019, p. 238). This limits women's access to technology, making it difficult to purchase or seek the services of tractors, processors, transportation, or labour (Duncan, 2004, p. 69). In addition to limited access to credit and technology, the USAID Ghana Gender Analysis identifies 'asymmetries in negotiating fair prices' and 'low participation in agri-business' (Britt et al., 2020). These constraints affect women's economic activities and create gender gaps in labour force participation.

Given decades of attention to gender equality in international policymaking, why do these inequalities persist? In this chapter, we argue that there are substantial gender gaps not only in agricultural practices, but also in agricultural policy processes. With increasing literature on the inequalities between women and men, understanding how women farmers participate in policy processes and how gender mainstreaming is implemented is relevant to explaining the persistence of gender gaps.

The chapter is based on research undertaken by DEMETER on gender equality and food security outcomes of land and agricultural commercialization, policymaking, and implementation processes. It analyses processes of agricultural policymaking, with particular emphasis on FASDEP II, the second iteration of Ghana's Food and Agriculture Sector Development Policy, initiated in 2008,[1] and Poverty Reduction Strategy Papers (PRSP). It also examines strategies used to mainstream gender in agricultural policy, and possible gaps therein. It draws on 39 interviews with stakeholders from academia, civil society organizations, development partners, the government, trade union organizations, and associations of farmer groups in various agricultural sub-sectors. Data was transcribed and thematic analysis was conducted using Nvivo.

The chapter is organized as follows: Section II presents the literature review and theoretical framework that underpins the study. It provides a brief on the push for the institutionalization of gender mainstreaming in organizations with the aim to bring about change in decision-making processes. The section also discusses Arnstein's theory of participation and how it situates with gender and policymaking in Ghana. Section III gives an overview of gender equality policies in Ghana and discusses gender mainstreaming as a transformative process to achieve gender equality and empower women in the agricultural sector. Section IV analyses the policy process in the formulation and implementation of FASDEP II to identify gender biases and how gender is addressed. It discusses processes that produce exclusions at the national and local levels and looks at some institutional challenges.

Literature review and theoretical framework

The UN Economic and Social Council (1997 states that:

> Gender mainstreaming is the process of assessing the implications for women and men of any planned action, including legislation, policies and programs, in all areas and levels. It is a strategy for making women and men's concerns and experiences an integral dimension of the design, implementation, monitoring and evaluation of policies and programs in political, economic and societal spheres so that inequality is not perpetuated.
>
> (UN, 1996)

Gender mainstreaming is, thus, tied to institutionalization (World Bank et al., 2009). Conferences during the United Nations Decade for Women pushed for the institutionalization of gender considerations in international and national agencies, arguing that social justice and equity for women would be more effectively achieved if linked to mainstream development concerns.

Advancing women's participation in decision-making was a central aspect of this strategy. Thus, Jahan advocated for what she called an 'agenda-setting approach' that requires giving women the opportunity to participate in policies that affect them (Jahan, 1996). She contrasted this with an 'integrationist approach', that merely adds gender issues to existing arrangements without challenging hierarchical gender roles and relations between women and men (Jahan, 1996; Reeves & Baden, 2002). But by seeking transformation through integration, gender mainstreaming 'encapsulates many of the tensions and dilemmas in feminist theory and practice' (Walby, 2005, p. 321) and offers different perspectives on how gender should be incorporated in development agendas (Mazey, 2000; Behning & Pascual, 2001; Verloo, 2001; Walby, 2001; Woodward, 2003).

Gender mainstreaming has been widely promoted by organizations that have analysed it as a change process with three elements: a vision and results (rights and power), a transformative technical planning process, and a commitment (Sida, 1997;). In this context, it is conceptualized in a theory of change that focuses on organizational culture, processes, and structures that aim at transforming policymaking (Daly, 2005). However, Cornwall and

Rivas (2015) argue that it often has taken the form of an 'add women and stir' approach. To them, genuine inclusiveness is not about giving people a say by 'inserting women into spaces created by others' but about ensuring that they are heard. A policy of gender mainstreaming thus needs to be paired with the genuine inclusion of women into politics.

Studies show that society in general is well secured and economically sound when gender equality issues are given adequate attention (World Economic Forum, 2017). It is thus fundamental to ensure that women have equal opportunities to participate meaningfully in decision-making. This includes recognizing intra-gender differences and that factors such as class, religion, ethnicity, or political affiliation can divide women. Most institutional structures are deeply rooted in norms and practices that are themselves rigid to change and serve as a platform for discrimination. Particularly important for this study are schedules that do not consider women's domestic roles, the means of communication, and long hours of sitting. When gender mainstreaming is well integrated in institutions, gender equality and women's empowerment would be seen not as a women's issue, but as transforming the development agenda.

In Ghana, scholars have noted with deep concern the low level of women's participation in policymaking. For instance, Kpessa and Atuguba (2013) critique the top-down approach of the constitutional review process as having been hijacked by 'educated, urban, and relatively organized Ghanaians over their underprivileged and traditionally marginalized counterparts, especially women' (Kpessa & Atuguba, 2013, p. 99). Yet the Affirmative Action policy target of achieving 40% women's representation in politics and decision-making has not been achieved (MoGCSP, 2015).

Arnstein's (1969) ladder of participation is useful for conceptualizing women's participation in policy processes in Ghana. Arnstein understands citizen participation as a redistribution of power that enables those excluded from political and economic processes to be deliberately included in decision-making and resource allocation. Arnstein's ladder consists of eight stages that indicate the degree of citizen's power ranging from non-participation to citizen control. The discussion in this chapter focuses on the intermediary stages, which are 'informing', 'consultation', and 'placation.'[2] According to Arnstein, these constitute progress to levels of 'tokenism' that allow the less privileged to be heard and have a voice, which will be submitted as full participation. However, this does not necessarily guarantee positive change in the situation of the powerless. Though the underprivileged can give advice, participation is tokenistic as long as final decisions lie in the hands of the power holders.

Under tokenism, *information flow* is usually one-way, from officials to citizens (top-down) with no channel for feedback or negotiation. Arnstein notes that in situations when information is provided at a late stage in planning, people have little opportunity to contribute for their benefit. She observes that such one-way communications are often via the news media, pamphlets, posters, and responses to inquiries (Arnstein, 1969). With regard to *consultation*, the common approaches are meetings and public hearings. Nonetheless, when participation is limited to any of these approaches, it becomes what she describes as a 'window-dressing ritual' or perceiving people as 'statistical abstractions', where attendance becomes the unit for measuring participation and a card for power holders to use as evidence of involvement. *Placation* is also seen as tokenism but with some level of participation that is limited to selected people in the community. In this situation, power holders choose representatives either because of their status or financial standing in the community. In many cases, the power to influence still lies in the hands of the powerful as they occupy majority seats and can therefore outvote the minority. Arnstein (1969) adds that factors such as the 'quality of technical assistance for articulating their priorities and the extent to which the community has been organized are important in determining the degree to which citizens are placated'. Does Ghanaian women's involvement in policy processes emerge as tokenistic, as Arnstein's theory postulates? That is, are they able to participate meaningfully, that is, in a way that makes a difference? In this chapter, we seek to answer these questions by looking at agricultural policy processes.

Gender mainstreaming and agricultural policy in Ghana: an overview

The Ghanaian government established the National Council on Women and Development (NCWD) in the aftermath of the first United Nations Conference on Women in 1975. Twenty years later, and in response to the call for gender mainstreaming at the 1995 Beijing Conference, the NCWD formulated guidelines for the promotion of gender equality, rights, and opportunities for women in Ghana (MoGCSP, 2015). This provided opportunity for enhanced cooperation between state and non-state agencies. Gender desk offices (GDOs) were created in ministries, departments, and agencies to ensure that policies and programmes incorporated gender concerns, and a full gender ministry, the Ministry of Women and Children's Affairs (MOWAC) was established in 2001. The Ministry spearheaded the development of the National Gender and Children Policy in 2004 with the goal of mainstreaming gender concerns into national development processes. MOWAC became the Ministry of Gender, Children, and Social Protection (MoGCSP) in 2013, and developed a more comprehensive national gender policy. The 2015 National Gender Policy provides the institutional framework and strategies towards achieving gender equality and women's empowerment targets with a national vision of 'a stable, united, inclusive and prosperous country with opportunities for all'. It continues to focus on 'mainstreaming gender and women's empowerment into Ghana's development efforts' at all levels of governance, and serves as a guide for all stakeholders to understand the strategies for addressing gender concerns (MoGCSP, 2015). The policy spells out five broad objectives: rights and access to justice, leadership and accountable governance, empowerment, promoting gender roles, and improving economic opportunities including engendering macro-economic and trade policies. In the areas of food and agriculture, it seeks to review and advance land reform and enforce the implementation of extension services to ensure gender equality and women's empowerment.

The policy is also being translated in the Ministry of Food and Agriculture. MoFA has focused on women in agriculture since the 1970s through the Women in Agricultural Development Directorate (WIAD). However, this did not entail a gender mainstreaming gender strategy. Thus, the Economic Recovery Programme from 1980 to 1983, and the Medium-Term Agricultural Development Strategy (MTADS) failed to mainstream gender and address women's issues, leading to the unequal allocation of resources in favour of men and failure in attaining the desired impact of reducing poverty particularly among women farmers (World Bank, 2009). WIAD developed the Gender and Agricultural Development Strategy (GADS) in 1997 to ensure the incorporation of women into agriculture and food security (Opare & Wrigley-Asante, 2008). The strategy was operationalized in 2004 and updated in 2015. Areas that GADS highlights include improved access to extension services, better access to credit, development of new processed products, and increased access and control over land.

GADS has been the main framework to institutionalize gender in the agriculture sector with the aim to address gender equity concerns (JICA, 2013). However, the 2002 Food and Agriculture Sector Development Policy, FASDEP I, was silent on gender. A poverty and social impact analysis (PSIA) conducted on FASDEP I failed to identify the lack of gender concerns in the document. FASDEP II sought to ameliorate the weaknesses in FASDEP I through extensive stakeholder consultations. But it continued to pay little attention to gender, which now appeared as a 'cross-cutting issue', together with land. Although consultations were held in each district of Ghana's ten regions, women's participation amounted to tokenism, as we show in this chapter. A range of policies following FASDEP II failed to put in place measures to address gender inequalities, including the 2013 National Seed Policy and the 2013 Fertilizer Policy. FASDEP II and the Medium-Term Agriculture Sector Investment Plan (METASIP) (2011–2015) emphasized gender mainstreaming mostly by addressing sector constraints on productivity, market access, sustainable production, and institutional coordination (MoFA, 2010). A key issue identified was the need to train staff at various units in MoFA as the workforce was dominated by about 73% men (Dittoh et al., 2015). But there was a lack of resources and implementation (MoFA, 2015). Given the complexity of challenges, there was little progress in closing gender gaps. The 2015 National Gender Policy seeks to reinvigorate these efforts. It aims to enhance the institutional capacity of MoFA, foster the production of disaggregated data, and improve extension service delivery, access to financial services, and information on land rights. The further development of appropriate technologies and new processed products, and environmental protection, are also on the agenda. Other agricultural programmes, such as the Global Food Security Strategy (GFSS), share similar goals and objectives.

In Ghana, many development partners, NGOs, and CSOs are involved in agriculture through funding, and participate in policy formulation and implementation. Such actors are very influential due to their financial power, negotiating skills, and strong research ability. They have been particularly active in advocating for gender mainstreaming, providing funding, and technical support. For example, the Ghana Country Plan (2018–2022), developed by United States Global Hunger and Food Security Initiative, explicitly focuses on gender and food security and aims to engage women and youth in agriculture-led growth (US Government Agencies, 2018). Often, inclusion of gender concerns becomes a condition for funding without which proposals can be delayed or declined (Dittoh et al., 2015). In a study conducted in the Upper East region of Ghana, gender units interviewed described programmes supported by donors as 'very active' and those supported by government as 'fair' (Dittoh et al., 2015). Similarly, NGOs and some CSOs were described as very active in gender advocacy and women's empowerment in this study. However, donors sometimes lobby for policies that suit their agenda and can be selective in funding only government policies that meet their goals, thus compromising on gender issues. The increasing number of actors with competing interests compounds the process, which may explain the loopholes in gender mainstreaming in part.

Making FASDEP II: women's participation and gender gaps

This section analyses the policy process to identify gender biases and how gender is addressed in the making of FASDEP II. It assesses the extent to which women's inclusion has been tokenistic, precluding meaningful participation in policy formulation. It discusses processes that produced exclusions at the national and local levels and looks at some institutional challenges.

The MoFA has the authority to lead in policy formulation in agriculture. The development of FASDEP II was a five-stage process. The first stage, led by a consultant from the University of Ghana, entailed identifying issues emanating from the implementation of FASDEP I. It involved institutions at the national level and the constitution of various committees based on profession and the linkage to agriculture, including a committee on gender and social welfare. Each committee produced a report digesting the problem, differentiating it from symptoms and proposing solutions, which the consultant put together to produce the zero draft of the policy. The second stage of the process involved nationwide regional consultation. An electronic copy of the zero draft was sent to stakeholders (such as CSOs and government agencies) for their inputs. The process involved nationwide regional consultations. The third stage was the organization of a national workshop where inputs were incorporated, and a final copy produced. The fourth stage was a presentation of this copy to the cabinet by the minister of food and agriculture. After studying the document, the cabinet asked questions about topics such as the financial implications of the policy and whether it addressed the needs of beneficiaries, before making a final decision. In a fifth step, the policy was then published and launched.

The development of FASDEP II thus provided multiple opportunities for information, consultation, and the inclusion of some. In Arnstein's words, it was planned as tokenistic participation that stopped short of true partnership, delegated power, and citizen control. However, as we will show, women were excluded even from tokenistic participation.

Processes of inclusion and exclusion: national level

Analysis of the policy process in the development of FASDEP II shows elements of a top-down approach and poor consultation of key stakeholders. To begin with, there was little gender mainstreaming or women's inclusion in the problem identification stage. In an interview, advocates such as the director of Women in Agricultural Development (WIAD), the director of the Peasant Farmers Association (PFA),[3] and a representative from the Department of Social Welfare and Development[4] indicated that they were not invited to the consultative meeting. If Arnstein's ladder of participation is anything to go by, then it is safe to say there was no participation of advocates for gender equality and peasants at this problem identification. For example, the WIAD director told us:

> No, I was not invited to the problem identification stage. They sent me the draft for my inputs (consultation), and reviewing documents and making inputs does not necessarily mean that the inputs will be accepted.
> (WIAD Director, Accra, 02/09/2016)

This response directly links to Arnstein's point on tokenism where information flow is one-way from officials to citizens (top-down), and while there is a channel for feedback, there is no room for negotiation. The less privileged are allowed to have a voice and evidence of this is submitted as full participation. However, soliciting for citizens' views provides no assurance that people have been heard; such participation is a mere formality.

In our interview, the consultant in charge of drafting the policy was very clear that there was no conscious effort to invite women's groups. However, she said the department of women in social welfare was represented, and it was expected that representatives of this institution would bring up gender issues. The assumption that social welfare can efficiently contribute to agricultural policy to address women farmers' concerns means a lot needs to be done. And while peasant farmers and gender equality advocates did not have the ear of the consultant, agribusiness, and the large national farmers' associations apparently did. According to the PFA director, agricultural input dealers were invited to a separate meeting for their views before the draft was circulated for comments. And in our interviews with officials at Ghana Statistical Service, they insisted that all key stakeholders were involved:

> They don't put it on their chest alone; they deal with all the players on the ground to actually get their policies right. I know we have the Ghana National Association of Farmers and Fishermen; and a lot of farmer-based organizations that government will always consult to do their policies.
>
> (Ghana Statistical Service, Accra, 06/09/2016)

Our government interviewees presented a seemingly ideal process, which in reality fell short. Thus, the WIAD director indicated that gender issues were considered in all the processes; however, a request for a list of women's representatives at the various stages was not available. And although the representative at the department of women in social welfare was expected to speak for women, she said she was never invited to the problem identification stage but was rather given a draft to review. Apparently, decision makers view gender mainstreaming as a matter of reviewing documents with no assurance that their feedback will be taken up. As posited by Arnstein, they have a voice but are not heard.

Our interview with the director of the Ministry of Lands and Natural Resources (MLNRs) was telling in this regard. Like other government staff, he also insisted on the Ministry's commitment to gender mainstreamed consultation, but explained why comments do not make it into policies.

> We do a lot of engagements through workshops, through various meetings and gender desk officers are invited. But the quality of the final document will depend on what comments and how strong the comments were made to reflect what is required. But at least on our part as government, at least from the Ministry of Lands and Natural Resources, I think that we really do quite a bit in trying to mainstream this.
>
> (Director, MLNRs, Accra, 02/09/2016)

The problem for this director was 'how strong the comments were made', which would require opportunities for getting across the message. The existing land policy does not explicitly refer to gender equality, no doubt linked to the tokenistic participation of gender advocates and a lack of opportunities to forcefully raise the issue and be heard.

Processes of inclusion and exclusion: local level

Participation and associated exclusions take a different form at the local level. The understanding that emerged from our interviews is that invitations for input are extended to the local level and these are included in policy development eventually. However, opinions were divided as to the effectiveness of the involvement. Various problems stood out in the consultation processes around FASDEP II.

To begin with, membership in farmer groups is an important means through which large numbers of farmers can be reached. The broad view was that there are a large number of farmer groups and farmers are ordinarily involved in policymaking. But the groups are highly fragmented with varying interest, and this is a challenge for presenting a united front. At Radio Gold,[5] it was noted that:

> The high number of farmers makes it difficult to capture all of them or most of them under the same umbrella. The lack of unity amongst the farmers makes the policy process difficult. The only challenge is that they themselves are not united. They are so many that it is very difficult working with them.
>
> (Interview at Radio Gold, Accra, 09/03/2016)

At the same time farmers, especially women, are sometimes excluded because they are not part of any farmer groups or associations. This makes it easy to sideline them in the process of policymaking. Even if women are integrated into participatory exercises, cultural values may become an obstacle. For instance, in parts of Ghana, women do sit with men to deliberate over issues. But sometimes those less educated are perceived not to have anything positive to add to decision-making; given that women in both rural and urban areas continue to be the least educated, their contributions are, thus, often valued less.

The setting of consultations can provide another obstacle. The director of PFA suggested that the environment at meetings can be intimidating. This is particularly important in local contexts where women are not allowed to sit with men. When a platform is created including top officials, such as CEOs, directors or their deputies, the discussions or the language used may be difficult for the local person to understand, and it may be difficult to make contributions. In the end, their presence just serves to document that they were involved in the process, but in effect, they were not able to make an input or articulate their concerns. This situation resonates with what Arnstein describes as 'window-dressing ritual' or perceiving people as 'statistical abstractions' where attendance becomes the unit for measuring participation, which power holders use as evidence of involvement.

In our interviews, participants highlighted a number of problems relating to the practicalities of consultations. They included the wrong timing of meetings, sending huge volumes of documents to be reviewed just a few days before such meetings, and a lack of resources to implement programmes. The following excerpt from an interview with NETRIGHT, a coalition of civil society organizations and individuals that promote women's rights and advocate for policy change in Ghana, illustrates the problem.

> People can be excluded directly or indirectly. For example, when consultative meetings are scheduled at a time most women farmers have gone to the farm they cannot attend, again if meetings are scheduled on a market day, women will not come. Apart from that we, those who advocate for women, are usually given a big document to read and provide inputs just a day or few hours to the meeting, which makes it impossible to effectively give inputs. Another issue is for example, using emails as the only means of communication and given the illiteracy level of women farmers, it is obvious that majority of them would be excluded because many do not have the know-how or access to email. … Despite these challenges, it will still be documented that the process was all inclusive just because an invitation was sent out.
>
> (Netright, Accra, 02/09/2016)

Participatory processes, thus, do not guarantee the inclusion of women or gender integration in policies. In the cases of poverty reduction strategy papers and FASDEP II, two policies we discussed with our interviewees, drafters have scarcely integrated participatory inputs. These findings invite a critical review of participatory policy processes to make sure they are not mere tick box exercises.

A final issue that emerged from our interviews pertains to a lack of transparency that stands in the way of full and effective participation. In the words of the director of PFA,

> Decision-making has not been transparent and there is lack of information because there are a lot of things happening in the villages that farmers are not aware of. No awareness creation, nothing. Normally, those of us who haven't been to school don't know about the laws so we're not part of it. When they are making the laws, they don't include us.
>
> (Peasant Farmers Association, Accra, 2 September 2016)

The issue reverts to inequalities derived from differential access to information. As Kpessa (2011) argues in his study of government and citizens' engagement and its interaction with policymaking, the current policymaking structure is fraught with challenges especially when the rural areas are disorganized, making elite capture of the process more pronounced.

Institutional challenges

Gender mainstreaming not only seeks the inclusion of gender considerations into policies through women's participation but, as outlined in Section II, also involves a strategy of changing institutions so that gender is taken into account in all stages of policy formulation and implementation and at all levels of government. This requires building institutional capacity, including gender expertise, and it requires resources.

In our interviews, we found a dearth of gender expertise. On the one hand, some interviewees recognized that this expertise was needed. For example, the director in the MLNRs talked about plans to recruit a social development expert in the context of the land administration project 'and her role purely is to ensure that issues of gender marginalization and all those things are dealt with' (Director, Ministry of Lands and Natural Resources, Accra, 2 September 2016). On the other hand, throughout the country, the Lands Commission has only two women at the regional level, the Northern and Upper East regions, and our research participants opined that, administratively, there is no gender consideration in top management positions.

An associated problem was the assumption that every woman in a leadership position is a gender specialist (Kunz & Prügl, 2019). According to WIAD staff, the appointment of WIAD directors has always been based on years of service, and in the past, long-serving women from the Plant Protection and Regulatory Services Department (PPRSD) were appointed to head the division. This 'add women and stir' approach (Cornwall & Rivas, 2015) entails a limited understanding of inclusion and participation and disregards the need for gender expertise. Our interview with Netright offers a vivid illustration of the problem:

> Gender mainstreaming is not effective because I know in some District Assemblies the gender desk is given to any woman at the discretion of the District Chief Executive. The position has nothing to do with the person's gender background, nor her interest in gender issues. At best, it becomes an added responsibility for the woman in charge of social welfare. I recall a case where the gender desk officer in the district was doubling as social welfare officer and we were running a programme on gender in the district for which we needed to work together. We never met the gender desk officer, she never participated in any of the programmes. I also noticed that they are not in the decision-making space; so even if they have ideas, it doesn't go anywhere
>
> (Netright, Accra, 02/09/2016)

The problem here emerges not only as one of expertise, but also of gender tasks added to existing responsibilities and gender officers lacking authority.

In addition to a dearth of gender expertise, respondents noted that resource constraints and technical challenges hindered implementation even when gender was incorporated in policy documents. Makinde (2005) argues that when there are resource constraints, people need to use their discretional powers to prioritize. In the process, conflicting interests often skew priorities to the disadvantage of women. A related limitation comes from the fact that the government prioritizes some sectors of agriculture over others. For example, the influence of the government is visible in the cocoa sub-sector but diminishes in the food crop sub-sector where women dominate. Unlike the cocoa sector, where the government regulates marketing and pricing, the food crop market is open and left in the hands of aggregators who control and negotiate prices to their advantage.

Limited government resources have created a vacuum for more resourced actors to take up policy implementation, which can affect gender inclusion. As Ankrah et al. document in this volume, private actors and international donors provide grants and loans, which leverages their increasing prominence in policymaking and implementation. Indeed, one of our interviewees suggested that most gender-related policies are sustained and functional because of donor support (Ministry of Land and Natural Resources, Accra). Interviewees were critical of this donor dependency because they recognized that partners would support policies only if they fell within their interest areas and tied with their goals. Additionally, well-resourced donors are able to invest in research that provides them with rich evidence to support their arguments and lobby for policies that are in their interest. While there is ample rhetoric about partnerships in development, the powerful role of donors raises the issue of women's participation in a new way. In Arnstein's theory, partnership goes beyond tokenistic inclusion and is most effective when locals are well organized, empowered, and well-resourced to support representatives. Gender mainstreaming thus cannot stop at including gender-specific goals into programmes and policies, but needs to help build the basis for meaningful participation. Without a stronger investment in expertise and resources, institutional strategies, such as gender mainstreaming, cannot live up to this expectation.

Minding the gender gaps – summary and conclusions

Do women participate meaningfully in policymaking and implementation in the agricultural sector in Ghana? The chapter has shown that though there are efforts to remove gender inequalities in the making of agricultural policy, this has not been effective, thus perpetuating the inequalities and gender gaps identified in many studies (Quisumbing et al., 1995; Udry, 1996; Makinde, 2005; Awumbila, 2006). A number of factors have been identified as sources of exclusion in the formulation and implementation of agricultural policy, with women, youth, the aged, and illiterates the most affected.

Although state organs have the authority in the policy process, the agriculture sector in Ghana involves many actors who play different roles and have diverse interests. Among them are the development partners, NGOs, CSOs, interest groups, and the private sector who dominate the policy space using strategies such as lobbying pressure or financial power. Current literature criticizes policymaking processes in Ghana for their overly elitist nature.

While it is widely acknowledged that many stakeholders are invited to the policymaking table, our respondents questioned the effectiveness of the involvement. At the national level, women's organizations were included in policy formulation in a tokenistic fashion. Excluded in the early phases, they were invited to comment on drafts, but their voices were perceived as weak. At the local level, respondents noted a lack of organization among women farmers, and cultural biases against their political participation. They also were concerned about the sometimes-intimidating setting of consultations, the wrong timing of consultative meetings, and the lack of lead-time given to local stakeholders to review huge volumes of documents just a few days before such meetings. Similarly, women's low educational levels and the expectation that they would not have anything positive to add to the policy process were used to justify their exclusion. Another challenge mentioned was a lack of data and a lack of transparency and awareness among farmers of the issues debated. Because of these difficulties, our respondents concluded that consultation processes were treated as a mere formality; the views of some participants would be taken for information, but they were rarely considered. In other words, the participation of stakeholders in agricultural policymaking was tokenistic, especially for women and other underprivileged groups.

In conclusion, gender inequalities continue to dominate in agriculture. This is reinforced by cultural tenets that reproduce them despite existing policies such as GADS, and despite superficial efforts to mainstream gender in FASDEP II and in poverty reduction strategy papers. It is clear that the policy formulation process and implementation reproduce structures of power that leave in place gender gaps. Agriculture is a critical sector that can help reduce poverty and thus achieve one of the most important development goals for the country. For this to happen, there is the need for careful analysis of gender issues to ensure proper mainstreaming that includes all groups of farmers in decision-making in a meaningful way.

Notes

1 FASDEP II is soon to be replaced with FASDEP III, which is in line with the UN Sustainable Development Goals and ECOWAS Agricultural Policy (Ghana News Agency, 2020).
2 The eight stages are: *manipulation, therapy, informing, consultation, placation, partnership, delegated power, and citizen control.*
3 The PFA is the apex farmer-based NGO in Ghana with a mandate to advocate for pro-poor agriculture and trade policies and other issues that affect the livelihoods of smallholder farmers. It focuses on increasing agricultural production, processing, and marketing through strengthening the capacities of farmers in policy advocacy. Though it includes both women and men, PFA has a strong focus on women farmers.
4 The Department of Social Welfare and Development is a public agency under the Ministry of Gender, Children and Social Protection (MoGCSP), with the responsibility of taking the lead role in integrating the disadvantaged, vulnerable, persons with disabilities, and the excluded into mainstream society.
5 Radio Gold is a local and privately owned radio station in Accra, the capital of Ghana.

References

Apusigah, A. A. (2009). 'The gendered politics of farm household production and the shaping of women's livelihoods in Northern Ghana.' *Feminist Africa*, 12(2), 51–68.

Arnstein, S. R. (1969). 'A ladder of citizen participation.' *Journal of the American Institute of Planners*, 35(4), 216–224, https://doi.org/10.1080/01944366908977225

Awumbila, M. (2006). 'Gender equality and poverty in Ghana: Implications for poverty reduction strategies.' *GeoJournal*, 67(2), 149–161.

Behning, U., and Amparo, S. P. (eds.) (2001). *Gender Mainstreaming in the European Employment Strategy*. Brussels: European Trade Union Institute.

Britt, C., Megan, I., Samuel, E., and Vivian, F. (2020). *USAID/Ghana Gender Analysis Report*. Analytical Report, Washington: Banya Global. Accessed May 16, 2021. https://banyanglobal.com/wp-content/uploads/2020/05/USAID-Ghana-Gender-Analysis-Report.pdf

Cornwall, A., and Rivas, A. M. (2015). 'From "gender equality" and "women's empowerment" to global justice: Reclaiming a transformative agenda for gender and development.' *Third World Quarterly*, 36(2), 396–415, https://doi.org/10.1080/01436597.2015.1013341

Daly, M. E. (2005). 'Gender mainstreaming in theory and practice.' *Social Politics: International Studies in Gender, State and Society*, 12(3), 433–450.

Dittoh, S., Snyder, K. A., and Lefore, N. (2015). Gender policies and implementation in agriculture, natural resources and poverty reduction: Case study of Ghana's Upper East Region. Colombo, Sri Lanka: International Water Management Institute (IWMI). CGIAR Research Program on Water, Land and Ecosystems (WLE). 22 p. (WLE Research for Development (R4D) Learning Series 3), https://doi.org/10.5337/2015.205

Duncan, B. A. (2004). *Women in Agriculture in Ghana*. Second. Accra: Friedrich Ebert Foundation.

FAO (2011). *The State of Food and Agriculture*. Rome: FAO

Ghana News Agency (2020). NGO advocates inclusion of agroecology in FASDEP III. Published November 29, 2020. https://newsghana.com.gh/ngo-advocates-inclusion-of-agroecology-in-fasdep-iii/

GSS (2019). *Ghana Living Standard Survey* 7. Statistical Report, Accra: Ghana Statistical Service. Accessed May 16, 2021. https://www.statsghana.gov.gh/gssmain/fileUpload/pressrelease/GLSS7%20MAIN%20REPORT_FINAL.pdf

Huyer, S. (2016). 'Closing the gender gap in agriculture.' *Gender Technology and Development*, 20(2), 105–116. https://doi.org/10.1177/0971852416643872

Jahan, R. (1996). 'The elusive agenda: Mainstreaming women in development.' *The Pakistan Development Review*, 35(4), 825–834, Islamabad. https://www.jstor.org/stable/41260001?seq=1#metadata_info_tab_contents (Retrieved 3/8/2020)

Japan International Cooperation Agency (2013). Country gender profile: Republic of Ghana final report.

Kpessa, M. W. (2011). 'The politics of public policy in Ghana: From closed circuit bureaucrats to citizen engagement.' *Journal of Developing Societies*, 27(1), 29–56.

Kpessa, W. M., and Atuguba, A. R. (2013). 'Grounding with the people: Participatory policy making in the context of constitution review in Ghana.' *Journal of Politics and Law*, 6(1), 99–110.

Kunz, R., and Prügl, E. (2019). 'Introduction: Gender experts and gender expertise.' *European Journal of Politics and Gender*, 2(1), 3–21, https://doi.org/10.1332/251510819X15471289106077

Makinde, T. (2005). 'Problems of policy implementation in developing nations: The nigeria experience.' *Journal of Social Sciences*, 11(1), 63–69.

Mazey, S. (2000). 'Introduction: Integrating gender – intellectual and "real world" mainstreaming.' *Journal of European Public Policy*, 7(3), 333–345.

MoFA (2010). *Medium Term Agriculture Sector Investment Plan (2011–2015).* Accra: Ministry of Food and Agriculture. Accessed May 16, 2021. http://www.fao.org/faolex/results/details/en/c/LEX-FAOC144958

MoFA (2015). *Gender and Agricultural Development II.* Strategy Document, Accra: Ministry of Food and Agriculture. Accessed May 13, 2021. http://www.fao.org/faolex/results/details/en/c/LEX-FAOC144957/

MoGCSP (2015). *National Gender Policy.* Accra: Ministry of Gender, Children and Social Protection.

Network for Women's Rights in Ghana. (2014). Report of desktop study on Gender Equality and Social Inclusion issues relating to the Land Sector in Ghana. Netright; https://www.star-ghana.org/learning-2/publications-and-resources/reports/137-netright-study-on-gender-equality-and-social-inclusion-iss

Opare, J., and Wrigley-Asante, C. (2008). *Assessment of the Gender and Agriculture Development Strategy (GADS).* Consultant report. Ghana: Ghana SASH.

Reeves, H., and Baden, S. (2000). Gender and development: Frequently asked questions. Bridge Development –Gender, (57), IDS, UK.

Sida, (1997). *Sida's Action Program for Promoting Equality between Women and Men in Partner Countries: Policy, Experience Analysis, Action Plan.* Stockholm: Sida.

Tsikata, D. (2016). 'Gender, land tenure and agrarian production systems in Sub-Saharan Africa.' *Agrarian South: Journal of Political Economy*, 5(1) 1–19.

Udry, C. (1996). 'Gender, agricultural production, and the theory of the household.' *Journal of Political Economy*, 104(5), 1010–1046.

United Nations (1979). Convention on the elimination of all forms of discrimination against women. United Nations.

United Nations (1996). *Beijing Declaration and Platform for Action. Adopted by Fourth World Conference on Women.* New York: United Nations.

US Government Agencies (2018). *Global Food Security Strategy (GFSS) Ghana Country Plan.* Washington, DC: US Government. Accessed May 22, 2021. https://www.usaid.gov/sites/default/files/documents/1867/Ghana_Country_Plan_Public_Version_WS_Edits.pdf

Verloo, M. (2001). Another velvet revolution? gender mainstreaming and the politics of implementation. IWM working Paper No. 5/2001. Institute for Human Sciences, Vienna.

Walby, S. (2001). 'From community to coalition: The politics of recognition as the handmaiden of the politics of redistribution.' *Theory, Culture and Society*, 18(2–3), 113–135.

Walby, S. (2005). 'Gender mainstreaming: Productive tensions in theory and practice.' *Social Politics*, 12(3), 321–343, https://doi.org/10.1093/sp/jxi018

Woodford-Berger, P. (2007). 'Gender mainstreaming what is it (about) and should we continue doing it?' In Andrea Cornwall, Elizabeth Harrison and Ann Whiteheads (Eds.). *Feminisms, Contradictions, contestations and challenges in development.* London: Zed Books. 122–134.

Woodward, A. (2003). 'European gender mainstreaming: Promises and pitfalls of transformative policy.' *Review of Policy Research*, 20(1), 65–88.

World Bank (2009). *Gender in Agriculture Sourcebook*. Washington, DC: International Bank for Reconstruction and Development.
World Bank, FAO & IFAD (2009). *Gender in Agriculture Sourcebook*. The World Bank, Washington D.C.
World Economic Forum (2017). *Global Gender Gap Report*. https://www.weforum.org/reports/the-global-gender-gap-report-201

7 Agricultural commercialization and gender mainstreaming in decentralized Ghana

The politics of business

Daniel Adu Ankrah, Dzodzi Tsikata, and Fred Mawunyo Dzanku

Introduction

Agribusinesses have become salient actors in the development of rural economies. Most governments in the global south have invited them to become 'partners' to improve agricultural productivity and strengthen food security (SDG-2), and to do so with an eye towards gender equality (SDG-5). The government of Ghana designed and implemented several agricultural and gender equality policies spanning several decades, including policies focused on commercialization and gender mainstreaming. However, inadequate financial resources continue to stifle the implementation of such policies. In this context, private sector actors (businesses) have become crucial partners, providing adequate funding and expertise.

Policies of decentralization appear to have facilitated the increased role of the private sector. Ghana was among the first African countries to initiate decentralization as part of the public sector reforms promoted by the World Bank in the late 1980s to early 1990s. A key argument in favour of decentralization was to deepen democracy, reduce bureaucracy, and address fiscal challenges. However, in practice, decentralization has been fraught with challenges that have led to inadequate fiscal transfers and left local governments short of resources.

In this contribution, we show how shortcomings in fiscal decentralization open the door for business interests to become entangled in development policies in Ghana. We argue that agribusinesses shape agricultural commercialization and gender mainstreaming through a subtle policy capture where they embed business interest into local development processes. Specifically, they push commercialization and cash crops at the expense of food crops, and they push women's participation in agriculture and their industries. This is occasioned by the fiscal challenges (financial constraints/inadequate budgetary allocation) faced by sub-national agencies that incline agribusinesses with financial resources to fill such gaps through the implementation of agricultural policies that meet their interest.

DOI: 10.4324/9781003202004-10

We analyse business practices in two districts in Ghana, namely Asunafo North Municipal in the Ahafo Region and Kwaebibirem District in the Eastern Region. Cocoa production predominates in our study communities in Asunafo North, and there is a strong presence of Mondelēz International, one of the major multinational agri-food companies with the headquarters in the United States of America (USA). In Kwaebibirem, our study communities show oil palm as a major production activity with the presence of the Ghana Oil Palm Development Company (GODP), a member of the Siat Group of Belgium, and Serendipalm Company Limited (SCL), a sister company of the USA firm Dr. Bronner's, which produces certified organic palm oil.

The rest of the chapter is structured as follows. The next section presents a literature review and a review of policy processes at the sub-national level, and challenges associated with decentralization in Ghana. The third section gives an overview of the study methodology and district characteristics. The fourth section sheds light on decentralization in Ghana. The fifth section presents agribusinesses' interest and influence. There is a concluding chapter (Section 6) that provides a summary, policy recommendation, and concluding remarks.

Literature review: decentralization and business

International donors advocate decentralization, arguing the merits of 'participation', 'ownership', improved fiscal management, and reduced bureaucracy. Major strands of literature (Ayee, 2013; Conyers, 2007; Goel, Mazhar, Nelson, & Ram, 2017; Resnick, 2017; Riedl & Dickovick, 2014) have supported this understanding, suggesting that decentralization increases participation and accountability and that it enhances governance at the local level. Agyemang-Duah et al. (2018) further argued that fiscal decentralization can lead to poverty reduction if there is a high level of financial autonomy of the sub-national agencies coupled with realistic budgetary allocation from the central government. But policies of decentralization have run into problems. Failure has been explained, inter alia, by over-centralization of resources, limited transfers to sub-national government, narrow local revenue base, lack of local planning capacity, limited changes in legislation, and the absence of a meaningful local political process (Ayee, 2013; Mockshell & Birner, 2015; Ndegwa, 2002).

To gain a better understanding of decentralization, we examine the three types identified by Crawford (2004) and Robinson (2007). *Administrative decentralization* or *deconcentration* defines the re-location of branches of the central state to local areas, entailing a transfer of powers to locally based officials who remain part of, and upwardly accountable to, central government ministries and agencies. *Fiscal decentralization* involves the transfer of fiscal resources and revenue-generating powers, inclusive of authority over budgets and financial decisions, to either deconcentrated officials and/or central government appointees or to elected politicians (Smoke, 2003). *Political decentralization* or *democratic decentralization* or *devolution* (of power) entails the transfer of powers and resources to sub-national authorities (Manor, 1995).

This chapter focusses on fiscal decentralization: a form of decentralization that seeks to address differential developments and encourage fiscal autonomy. It is also supposed to achieve improved governance, greater accountability, improvements in revenue mobilization and the quality of local services, and contribute to macro-economic stability (Agyemang-Duah et al., 2018; Ankamah, 2012; Inanga & Osei-Wusu, 2004). Some of these goals appear to be reached. Ankamah (2012) indicated that fiscal decentralization promotes high-efficiency values where community members participate and hold government officials accountable in their financial management. Governance is essentially deepened through increased accountability; the local government becomes more responsive, and decision-making relating to financial allocation gets closer to the local peoples (Ankamah, 2012; Bawole, 2017; Crawford, 2008; Inanga & Osei-Wusu, 2004). But some counter-arguments discredit fiscal decentralization. Ankamah (2012) and Prud'Homme (1995) indicated that the local people lacked the political power to hold their leaders accountable; local politicians were inclined to be as corrupt as the national level politicians; fiscal decentralization worsened fiscal imbalances, and local governments lacked prudent public expenditure management systems. But none of these studies have addressed how constraints associated with fiscal decentralization interact with the enhanced role of the private sector in development, and how it has promoted the role of agribusinesses in the implementation of policies. The literature on businesses, neoliberal agricultural policies, and how they address gender inequality in the global south, appears not to favour women but rather worsens existing differential gender access and relations.

Lanz, Prügl, and Gerber (2020) showed how businesses take advantage of the neoliberal agricultural policies in superficially prosecuting gender equality. Their study showed how the Global Agro-Development Company (GADCO) entered a 'community-private partnership' agreement in the Fievie Traditional Area in the Volta Region of Ghana. This company failed to address existing gender inequalities without considering the intersectionality with power relations and local institutions that shape resource management. The company's approach constituted a reductionist one where women were essentially included to increase their numbers. This approach only privileges a few women of good class and or status and only serves as a cover-up for businesses' exploitative practices. The authors argued that neoliberal development approaches tended to underscore macro-economic development that advantaged the international and private businesses to the neglect of local power relations. Other strands of literature (Nyantakyi-Frimpong & Bezner Kerr, 2017; Prügl, 2015; Tsikata & Yaro, 2014) showed that large-scale land acquisitions for agricultural development and businesses worsened and often deepened gender inequality. On the contrary, Behrman, Meinzen-Dick, and

Quisumbing (2012) also illustrated the benefit of such acquisitions in the provision of employment, production of new crops, and inclusion in out-grower schemes for women. Prügl (2015) adduced evidence to show how transnational companies under the label of corporate social responsibility failed to empower women through the disguise of neoliberal capitalism in the case of India (South Asia). Tsikata and Yaro (2014) showed how large-scale land acquisition for agricultural projects in the Northern Region of Ghana failed to address gender differences and women empowerment in the case of the Integrated Tamale Fruit Company (ITFC). The authors argued that businesses could have a good model but if they failed to take into account existing differential gender differences and biases, it might eventually only limit women's access to opportunities and their empowerment thereof. And indeed, this was the case in the instance of the ITFC project in the Northern Region.

District characteristics and study methodology

Data for this study were collected in five communities – Asumura, Kasapin, and Goaso in the Asunafo North Municipal (Eastern Region) and Bomso and Abaam in the Kwaebibirem District (Ahafo Region). The Asunafo North Municipal has agriculture as the dominant (72%) sector that employs about 44.5% of individuals. Crop farming constitutes the predominant (97%) farming activity. Cocoa production is dominant among the crops sector. Oil palm, soap making, cassava processing, and limited cashew processing are the main agro-processing activities in the district (Ghanadistricts, 2021).

Cocoa production in the municipality attracted an international agro-processing company – Mondelēz International, an American cocoa buying company. The company has a net revenue of $26 billion with a presence in 150 countries.

Kwaebibirem District has a population mostly (57%) rural. Forty percent are engaged as skilled agricultural, forestry, and fishery workers. The majority (71%) of the households are engaged in agriculture. Crop farming is the dominant (98%) agricultural activity. There are two agribusinesses active in the district. The Ghana Oil Palm Development Company (GOPDC) is instrumental in commercializing oil palm production via out-grower schemes. Unlike GOPDC, SCL works with smallholder farmers that operate between two and three hectares. In this way, SCL accesses over 10,000 hectares of organic oil palm. The company deliberately engages farmers and wage labourers in labour-intensive activities as a matter of assuming social responsibility and creating jobs. With an annual production capacity of between 500 and 1,000 tonnes, the scale of production is small, which does not make the company economically viable. However, the founder, Dr. Bronner's, operates the business from a social enterprise perspective, seeking to deliberately create a market niche for organic certified palm oil regardless of its profitability. SCL builds the capacity of the main government sub-national agency – the Department of Agriculture – to train farmers in organic fair-trade production.

Serendipalm Company Limited remunerates out-grower farmers who engage as casual labourers at a higher rate. Generally, we observed that agribusinesses in both Kwaebibirem District and Asunafo North Municipal offer higher premiums and bonuses to farmers. This incentivizes farmers to be commercially oriented towards cash crops production.

Asunafo North and Kwaebibirem thus constitute interesting case districts to study the role of businesses in directing local development. Our research is based on data collected in 2016, following a reconnaissance visit in 2015. We conducted eight focus group discussions (FGDs), 15 key informant interviews (KIIs), and 15 policy interviews (PIs). Key informant interviews were particularly important for gauging the effects of decentralization and the role of businesses. They included PIs with members of the district assemblies and decentralized government agencies. In addition, we organized FGDs and interviews with residents in our five case study communities. Participants drawn for the FGDs were purposively selected based on the economic activities engaged in, age, and status (indigene versus migrant). Because we were interested in the effects of gender, we organized male and female FGDs. The FGDs included six to ten members and lasted for not more than one hour and 30 minutes. We divided our case communities into four equal parts based on the existing demarcations and purposely sampled respondents from four quadrants in each community. The data collection instruments focused on policy issues, such as how fiscal policies are designed, the actors involved and their roles, who influences what, who funds what, challenges encountered with fiscal decentralization at the sub-national agencies, business objectives, agribusinesses internal policies, policies implemented by agribusinesses and why, and which gender issues are embedded in agribusinesses policies and why, etc.

Decentralization in Ghana

Ghana's decentralization system appears to be relatively well established and robust (Resnick, 2017; Riedl & Dickovick, 2014). The Provisional National Defence Council (PNDC) regime led by Flight Lieutenant Jerry Rawlings introduced the system in 1988 by enacting the Local Government Law (PNDC Law 207). This created the district assembly (DA) as the key responsible institutions in 110 districts designated within the country's 16 administrative regions. The principles of decentralization and participation in government were emphasized greatly in the 1992 Constitution, and it was envisaged as the means to entrench democracy. The DA is the highest administrative and political body that exercises legislative and executive functions. The DA is followed next by a District Executive Committee (DEC), which functions as the executive and coordinating body. A government appointee chairs the DEC and functions as the political and administrative head (Ghanadistricts, 2021). Several sub-government agencies including the Department of Agriculture (DoA), Commission on Human Rights and Administrative Justice

(CHRAJ), National Commission for Civic Education (NCCE), Nutrition Department, Planning Department, and Office of Stool Lands fall under the DA. These government agencies tend to have collaborative roles existing among themselves in achieving broad sector objectives. In practice, there appears to be limited collaboration among these government agencies in the two study sites due to limited financial resource allocation to the respective agencies. This limited financial allocation affects effective collaboration between both government and other government local authority agencies as well as government and private sector agencies. This indicates the fiscal challenges associated with Ghana's decentralization efforts. Delays characterize central government financial disbursements to the sub-national level. Stronger synergies are consequently not fostered among the government agencies.

In our study sites in Asunafo North and Kwaebibirem, all the three types of decentralization categorized by Crawford (2004) and Robinson (2007) are practiced, but there is apathy in participation and incomplete fiscal devolution. Riedl and Dickovick (2010) indicate two-thirds of DA members are elected on individual, non-partisan criteria with one-third appointed by the central government and a DCE. This form of decentralization is in line with political decentralization. However, the independence of the elected members is contested. The demerit with the political decentralization is that the DCE or municipal chief executive (MCE) is nominated by the central government. Therefore, there is a high tendency for the selected DCE or MCE to be accountable to the central government rather than the local people. This situation defeats the major merit of accountability. Ribot, Agrawal, and Larson (2006) argued that 'if powers are decentralized to actors who are not accountable to their constituents or superior authorities, then decentralization is not likely to accomplish its stated aims'. This leads to incomplete devolution whereby the local assemblies rely on the central government. Jütting et al. (2005) indicated that Ghana's early decentralization experiences consisted of mere deconcentration of planning and administrative capacities without the actual transfer of power, leading to limited impact. In the Asunafo North Municipal Assembly, the membership of the assembly is made up of the MCE, one assembly member elected from each of the 25 electoral areas, an elected presiding member. In total, government appointees do not exceed 30% of the total membership of the assembly. The Executive Committee is chaired by the MCE and consists of not more than a two-thirds majority of the membership (Ghanadistricts, 2021). This membership is similar to what pertains in the Kwaebibirem District. The local government is expected to be more responsive than the central government, but this argument remains weakened given the high fiscal dependence on central government and private enterprises.

Ghana's decentralization fits with the administrative decentralization where main central government agencies can be found within the DA. The government agencies remain part of, and upwardly accountable to, the central government ministries. Conyers (2007) argued that devolution of powers to local elected authorities did not necessarily ensure addressing the local interests and needs. There is evidence suggesting that elected authorities have turned out to be less representative and responsive than central government. Cabral (2011) argued that the limited evidence available suggests, however, that decentralization is yet to deliver its promised outcomes.

Concerning fiscal decentralization, DAs have the power to mobilize funds through internally generated funds (IGF). This is done through taxes, permits, fees, incenses, and fines. The DAs also, to some extent, have authority over budgets as they are involved in drawing up a composite budget. Government financial allocations to decentralized agencies remain inadequate and plagued with undue delays. This provides an avenue for private sector actors including agribusiness with adequate financial resources to implement policies that align with their interests and objectives.

In the process of policy formulation at the sub-national level, voices must be heard. In terms of who gets heard in policy design and formulation, it is anticipated that broad representation of constituents will have equal participation and voice. This, however, appears elusive; rather, there is a domination by a few players who are well resourced especially in the implementation of agricultural policies. Farmer-based organizations (FBOs) indicate that their voices are not given the needed attention and value in policy discourses. Consequently, FBOs do not see themselves as a powerful force in influencing agricultural policy but rather perceive sub-national agencies such as the DoA to be responsible for influencing policy direction, asserting influence, and having agency. There are FBOs that are formed by the grassroots themselves (natural FBOs) and also the artificial FBOs that are formed by the DoA in implementing specific government and development partners' interventions. In most cases, artificial FBOs are dominant. This has implications for the FBOs' sustainability. This is because most artificial FBOs collapse after the expiration of a project intervention. Farmers in their individual capacity and as members of FBOs complained extensively about, for instance, the destruction caused by artisanal and small-scale mining (ASM) activities in the Kwaebibirem District but achieved little success in pushing for relevant state policies to protect the environment. In FGDs, participants indicated that:

> We have witnessed in our district the indiscriminate mining activities taking place that is rendering our lands non-arable and posing danger to residents, the community and the environment. We have complained bitterly about this hoping that something will be done but it seems that we are hoping against hope.
>
> (FGD, Asuom, 24 February 2016)

Similarly, in Asunafo North, farmers remained unsuccessful in negotiating for lands to be apportioned from the forest reserves against a backdrop of land scarcity.

> In our district, lands are not available and the only option that we will like our leaders to consider is to apportion some lands in the forest reserves to us to cultivate cocoa. This is because our municipal is well noted for cocoa production, we have done all that we can but to no avail. We however think that there are influential people who help some individuals to fall timber from the forest reserve illegally.
>
> (FGD, Asumura, 12 February 2016)

The statement suggests that grassroots voices received little attention while government appointees and individuals with connections can leverage influence. And they are actually able to implement desired changes.

The next section examines two agribusinesses (Serendipalm and Mondelēz International) to examine the extent to which business interests contribute to agricultural commercialization and gender mainstreaming.

Agribusiness influence and interests

Agribusiness financing constitutes a major force in shaping the implementation of agricultural policies in Ghana. The desperate need for funding was made clear in our interviews. Thus, the planning officer at the Asunafo Municipal Assembly told us:

> We have very beautiful plans but it's the implementation which is difficult because we don't have the resources. The government doesn't provide enough resources for its implementation. ... plans on agriculture, gender, and nutrition and all the good things that need to be done, that ... we do have. It is the resources to implement the plans that constitute the challenge; so at the end, we are not able to achieve the expected results.
>
> (NI, 15 February 2016)

In another interview at the national level, a government official similarly indicated that:

> we budgeted for GHC 10,000,000 but the Ministry of Finance gave us GHC 2,000,000; so how do we implement what we are supposed to implement? So that is where we are at the moment.
>
> (NI, 15 February 2016)

Agribusinesses help fill this funding gap and consequently greatly influence agricultural commercialization and gender mainstreaming. They do so through two pathways: they create their own policies and programmes, and they engage in agricultural extension work – sometimes riding on the back of the extension service of the DoA.

An example of a company seeking to assert its influence through its policies and programmes is Mondelēz International. In Asunafo North, the agribusiness promoted commercial agriculture through its Project on Environmental Sustainability and its Policy for Cocoa Production (ESP)/Mondelēz Cocoa Life Programme. The project worked through international non-governmental organizations (INGOs) and the government sub-national agencies (in a public–private partnership) in five core areas – livelihood, community development, youth development, and environment. It included an INGO (Care International), the United Nations Development Programme (UNDP), and COCOBOD, which collaborated to implement the ESP in eight districts, including Asunafo North. The objective of the ESP is to ensure environmentally friendly production practices, protect the forest against encroachment, improve farmer livelihoods, ensure sustainability in cocoa production, and implement climate change resilient practices. So far, the ESP has successfully trained 1,668 extension agents to provide capacity building on silviculture for carbon stock improvements, natural resource management, and sustainable ecosystem practices to an estimated 34,914 farmers, of whom 38% are females (UNDP, 2021).

Serendipalm similarly seeks to establish commercially viable and environmentally friendly practices by promoting sustainable organic oil palm production, and more recently organic cocoa, without inflicting climate-related hazards. Like Mondelēz International, it also provides funding in implementing the agricultural commercialization of cash crops. Specifically, the company provides a stable premium for crude palm oil (CPO) produced even though it is not economically viable. This is to essentially create a market niche for organic oil palm marketed in the United States. Both companies also have deliberate policies to increase women's direct participation in agricultural production activities, which tend to favour men. Essentially, fiscal challenges afford SCL and Mondelēz International the opportunity to shape and implement agricultural policies and promote agricultural commercialization by filling the fiscal gap.

A second way in which businesses influence local development is through participation in agricultural extension, which helps alleviate shortages in public resources. We observed instances where the DoA was co-opted to carry out its mandate by both SCL and Mondelēz International. Even though helping the DoA to carry out its mandate is not out of place, policy priorities that finally get implemented differ. In the instance cited above, Serendipalm is interested in organic production, this policy direction and implementation cannot be read-off linearly from the broad agricultural sector policy. Implicitly, the DoA might want to promote environmentally sound production practices but not specifically through organic production.

Given the fiscal challenges, the DoA co-implements the activities that generally address broad sector needs. In this regard, prioritization in terms of which activities get implemented might differ barring the financial constraints faced by the DoA. In Kwaebibirem, SCL supplies input to farmers, provides extension services, and secures access to markets. It works with the DoA's agricultural extension agents to carry out its core mandate. Cognizance needs to be taken of the fact that the DoA, even though it constitutes the extension conduit, would have otherwise been rendered limited in scope in the absence of the financial assistance provided by the agribusiness. In typical cases, the fiscal transfer is characterized by delays. For instance, fiscal transfer meant for the first quarter is often delivered during the third quarter. In an interview, a female casual worker with SCL indicated that:

> Serendipalm Company Limited has extension workers who visit our farms as well as convene meetings to teach us how to go about our farming activities in an environmentally friendly way. They tell us that there should be no rubber bags in our farms because the rubber destroys the soil.
>
> (PB, SCL, Asuom, 24 February 2016)

Serendipalm Company Limited and Mondelēz International thus deliver private extension services through the public medium extension service. Even though the policy framing and implementation appear participatory, the agenda-setting and implementation of agricultural policies are greatly shaped by these two agribusinesses. Eventually, the private sector has a greater hand in defining what constitutes good and sustainable farming practices.

Mondelēz International in Asunafo North, as a case in point, implements policies and activities through the public sector sub-national agency – the DoA. In an interview, a female cocoa farmer indicated that:

> We have a farmer group called the Asumura Cocoa Farmers' Society, we are taught to inter-plant the cocoa with plantain or cassava. Mondelēz through its Cocoa Life policy trained us on the time to plant cocoa and the recommended chemicals to apply using the public extension officers.
>
> (AA, Asumura, 12 February 2016)

It is through the company that the farmers learn what, when, and how to plant, and they are being locked into an input-intensive industrial production system to deliver products for an international market. There is a policy, Cocoa Life, generated through corporate procedures rather than local democratic processes.

Thus, what is implemented at the district level depends on who has money, and not on public policy. In effect, government policies that align with private sector policies are the ones that finally get implemented. This is not to suggest that the policies conflict, but essentially policies and activities that promote the direct interest of the private sector receive maximum financial assistance to get implemented. Conversely, public sector priority policies and activities do not get implemented by the private sector businesses if they do not align and inure directly to their benefits. Shortages of public funding thus constitute an avenue where private companies exert their influence in the implementation of agricultural policies. In the ensuing sub-section, we discuss such influence in the areas of agricultural commercialization and gender mainstreaming.

Agriculture commercialization

In this section, we show the effects of the financial influence that the two agribusinesses have in shaping agricultural commercialization. This is made manifest in the organic production of oil palm and the dedicated attention given to cocoa production. Given that cocoa and oil palm are cash crops, it points to the selective interest that agribusinesses have in cash crops relative to the food crops.

In Kwaebibirem District, the main private sector actors include the GOPDC and the President's Special Initiative on Oil Palm[1] (PSI on Oil Palm), and they constitute the major driving forces in shaping agricultural commercialization. Serendipalm Company Limited supports the commercial production of organic oil palm even though there appears to be no explicit policy on organic oil palm production in Ghana. Serendipalm Company Limited has, however, worked mainly through the DoA by engaging agricultural extension agents to provide technical and advisory support services to farmers.

In contrast to the palm oil sector in Kwaebibirem, cocoa production in Asunafo North includes a highly visible, and financially strong state actor – the Ghana COCOBOD.[2] COCOBOD has the needed financial resources to implement a lot of government policies that influence commercial cocoa production – for example, the provision of free hybrid cocoa seedlings and mass cocoa spraying exercises. However, even though farmers acknowledge COCOBOD efforts, most FBOs within the municipality associate more readily with Mondelēz International than with COCOBOD and the DoA. This conspicuous association is a result of the direct assistance (monetary and technical assistance) received from Mondelēz International through the DoA, COCOBOD, NGOs, and INGOs. There are multiple channels of influence by Mondelēz International and this has improved visibility.

A major boost is given to agricultural commercialization through input subsidies. Serendipalm Company Limited, for instance, incentivized farmers to produce commercial organic oil palm by making available oil palm seedlings to all of its employees who were interested and had farmlands as equity to enable them to cultivate a two-acre plantation and receive a guaranteed price. Employees willing to cultivate more than two acres were given oil palm seedlings to plant and deductions were made from their salaries. Deliberate effort was made to ensure that monies deducted at the source remained minimal to prevent hardship on the part of its employees. The deductions ranged between GhC10 and GhC30 monthly. A female casual employee with SCL indicated in an interview that:

> Serendipalm Company Limited gave us oil palm seedlings to plant on an organic basis to sell to the factory. Once we the workers can secure

> land, the factory gives us seedlings free to plant and supply the produce to them. Every worker has to look for his or her land and the factory gives the employee seedlings to plant on a 2 acre- land for free. Then when the worker wants more seedlings beside the 2-acres free seedlings provided, it is sold to the individual, and payment is deducted in instalments from the person's salary. Any time they pick new workers, they give them free seedlings for two (2) acres of land.
>
> (PB, SCL, Asuom, 24 February 2016)

The agribusinesses are attractive to farmers because they inject funds into the local economy. For example, Mondelēz International pays bonuses whenever production targets are met. In Asunafo North, Mondelēz International paid over US$4,000,000 per year for the 2015 farming season to farmers as bonuses for meeting production targets. Similarly, SCL guarantees a fixed price for fresh fruit bunches even though prices offered for CPO witness intense fluctuations on the international market. Dr. Bronners, the proprietor of SCL, purchases at a price fixed even though it appears not profitable. While social responsibility provides the motivation, this practice also constitutes a deliberate policy to incentivize the commercial production of organic oil palm. An officer who works at SCL indicated that:

> Dr. Bronners has been subsidizing our product for about $500,000 annually to keep us in production when we first started producing organic Crude Palm Oil (CPO) to keep us in production and at that time we were selling the most expensive CPO in the world and it was Dr. Bronners who was ready to buy from us and keep providing all the incentives we needed.
>
> (COB, Internal Control Manager, Serendipalm, Asuom, 24 February 2016)

While the injection of money is welcomed from the perspective of individual farmers, it also creates problems. First, this makes newly established agribusinesses with less capital to struggle and even collapse after a few months or years of operation. Unlike South America, where there are government policies on oil palm that protect the interests of all the domestic players and ensure an even playing field, there is an absence of a government policy on price regulation for CPO in Ghana. This implies that agribusinesses control the entry of new businesses through price regulation by rendering sub-national agencies redundant in regulating prices or providing a policy framework in this regard. This essentially limits the number of players who influence agricultural commercialization, and privileges financially well-resourced businesses. In the Kwaebibirem District, SCL and GOPDC set the price for crude oil palm contingent on the world market and what favours the businesses. The GOPDC in particular is known to raise its price for fresh fruit bunches whenever a new entrant offers a more competitive price.

Second, the kinds of incentives offered for cash crops are missing for non-cash crops such as maize, plantain, and vegetables, which are central to peoples' diets. Because they are sensitive to this critique, both SCL and Mondelēz International encourage farmers to dedicate lands for food crop production and encourage smallholders to diversify. Nevertheless, their main interest is in the commercial production of cocoa and oil palm, and greater attention is given to incentivize this. An officer of Mondelēz International spelt out the logic:

> Mondelēz International use cocoa for their production so whatever we do over here should be geared towards cocoa production else they will take less interest. I can tell you that if the farmers in Asunafo North Municipal can meet their target, they are going to get $4,000,000. One wonders if plantain or maize producers will get the same incentive. Thus, any rational human being will want to go into cocoa production. That is why more attention is paid to cash crops to the neglect of food crops.
>
> (JB, Goaso, 11 February 2016)

We conclude that the effects of agribusinesses in influencing agricultural commercialization are manifested in three ways. First, there is a deliberate development of organic CPO market niche on the international market (USA) with Ghana as a raw production base – a strategy that extends the colonial production practices. Second, there is a selective fostering of commercialization for cash crops (cocoa and oil palm). Third, there is the distortion of CPO prices to out-compete new entrants in the case of oil palm. With regard to cocoa, the Ghana COCOBOD is the only mandated organization that purchases all cocoa in Ghana, so the price distortions phenomenon is effectively addressed. Bonuses are, however, paid directly to farmers to encourage the meeting of production targets influencing the commercialization agenda. Finally, there is the crowding out of food crops in both cases relative to the promotion of cash crops.

Gender mainstreaming

In this section, we argue that agribusinesses shape the way gender is mainstreamed into agricultural development but fail to adequately address existing gender inequalities given the little attention paid to local institutions responsible for natural resource management and the powers they embed. However, agribusinesses are visibly seen implementing gender mainstreaming that targets increasing female participation in agricultural development and their industries different from the government's approach. This is due to financial challenges that confront government agencies.

Broad sector policies on gender mainstreaming exist in Ghana, spearheaded by the Women in Agriculture Development (WIAD), a unit of the Ministry of Food and Agriculture (MoFA). But at the sub-national level, we found that most of the gender mainstreaming policies directly implemented emanate from the two agribusinesses.

For instance, Mondelēz International and SCL both have in place gender mainstreaming policies that encourage more female participation in cocoa and oil palm production, which is dominated by male farmers. Mondelēz International facilitated women's increased participation in cocoa production through conscious efforts that helped women to become independent cocoa farmers. Indeed, the business encouraged women to approach cocoa production as a full-time business. Mondelēz promotes women's active participation in FBO activities and management. This is particularly so for the FBOs that Mondelēz works with within the Asunafo North Municipal. Deliberate efforts have increased (40%) female membership in cocoa FBOs. Women have been encouraged to take up leadership positions. A case in point is the Asunafo North Municipal Cooperative Cocoa Farmers and Marketing Union Limited, where a woman is the vice president of the group, with other executives within the group being women.

Similarly, SCL, at the inception of the company, started working with a women's group known as Danini Palm Oil Women Association and trained them on fair-trade production of organic CPO for export. But for SCL, women are not just farmers – they are also employees. This is different in the case of Mondelēz because cocoa is not processed locally but exported as raw cocoa beans. SCL deliberately hires more women in its production site than men, employing 130 women out of a total of 230 staff at the time of our interviews. Agro-processing activities, such as oil palm processing into palm oil, generally advantage women given the gender regimes and gender division of labour, since many of the tasks involved in the oil palm value chain activities are considered women's work. Additionally, there is a deliberate effort by management to employ more women because they are considered to be more meticulous than men and the tasks involved in the oil palm processes require a depth of detail. In an interview with a male employee of SCL, he indicated that:

> Serendipalm does not discriminate towards any particular gender because everyone is treated equally and we have the majority of our population being women because the men do the harvesting, pruning, cracking of the fruits and it requires only a few men but when it comes to what the women do, we need more of the women to do things like carrying FFB, picking loose fruit and cleaning fruits. Steaming and clarification are undertaken by both men and women.
>
> (CB, male, Asuom, 24 February 2016)

Aside the pre-existing gender divisions of labour in cash crop production, SCL deliberately provides opportunities for women in their production activities. They do so by employing more women in their agro-processing sites and encouraging women to engage in the cultivation of oil palm. Literature

(Lanz et al., 2020; Nyantakyi-Frimpong & Bezner Kerr, 2017; Prügl, 2015; Tsikata & Yaro, 2014) however suggests that the employment of cheap female labour in agro-processing activities is typical and does not translate into gender mainstreaming and empowerment but rather worsens existing gender differences.

We argue that businesses that are financially endowed influence agricultural policy implementation. It is interesting to note that a key government agency – COCOBOD – also has a policy on women and youth in cocoa. The experience is that women tend to be more responsive to training, so there is a deliberate policy to push more women to minimize the perception of marginalization. This resonates with Serendipalm's instrumental reasoning on women's inclusion. Both government and private sector actors recognize the need to mainstream gender in their operations. Here again, COCOBOD being more financially sound than its counterpart government agency (DoA) is seen implementing more policies in this regard. The need for adequate funding to facilitate the implementation of government policies simply cannot be over-emphasized. Overall, aside from COCOBOD, private sector businesses tend to be more financially endowed and dominant in the implementation of agricultural policies.

This is not to suggest that gender inequalities do not exist. Indeed, the agribusinesses' efforts recognize existing gender inequalities and go beyond seeking to empower individual women. A key gender inequality that cannot go unnoticed has to do with the differential access to productive resources, especially land. Ankrah, Freeman, and Afful (2020) confirmed differential gendered access to productive resources in their study in southern Ghana. In both districts, it emerged that female farmers in most cases had to go through an adult male to access farmland. The current land tenure arrangement and the lack of access to land hinder women from engaging in large-scale commercial cash crop production in both Asunafo North municipal Assembly and Kwaebibirem District. This leads to situations where women find themselves engaged in the cultivation of food crops considered to be 'female crops'. This finding confirms Lanz et al. (2020) that showed gender was just reduced to increased female participation but failed to address intersectional inequalities due to little attention paid to local institutions for resource management.

Agribusinesses have generally pushed for increased numbers in female participation in agriculture and their industries. But there appears to be little evidence of how this has addressed pre-existing gender inequalities and the power embedded in local institutions that shapes resource management. Particularly those that address the unequal gendered relations and power asymmetries.

Summary, conclusion, and policy recommendations

In summary, the study found that policy implementation at the subnational level is dependent on agribusinesses with financial resources that implement agricultural policies and activities that best suit their interests vis-à-vis local-level policy. Serendipalm Company Limited and Mondelēz International were found to be the major driving forces in influencing agricultural policy implementation at the sub-national level in the Kwaebibirem District and Asunafo North Municipal Assembly. This is seen in the influence exerted by the two agribusinesses in agricultural commercialization and gender mainstreaming. In contrast, the Ghana COCOBOD remained a major influence on cash crops policies even though it remained a state institution. This is because the Ghana COCOBOD is comparatively better endowed with financial resources than traditional sub-national agencies. We conclude that Ghana's decentralization is fraught with fiscal challenges. This has allowed agribusinesses with financial resources to implement most agricultural policies that suit their interests. And this immensely shapes agricultural commercialization and gender mainstreaming in southern Ghana.

Both the central government and sub-national agencies are encouraged to address major challenges associated with fiscal decentralization. This can be done by increasing budgetary allocation that meets the needs of sub-national agencies by getting innovative with national revenue mobilization. This will enable sub-national agencies to implement relevant agricultural and gender policies and help reverse the dominance of policy implementation by agribusinesses. Timely budgetary allocations to sub-national agencies are encouraged to prevent typical situations characterized by delays in implementing activities scheduled by sub-national agencies. Increased advocacy is needed to encourage women's representation in local-level governance and to address challenges posed to women involved in agricultural production with regard to their access to productive resources. Agribusinesses should go beyond increasing numbers of female participation in agriculture and their businesses and pay adequate attention to local institutions responsible for natural resource management and the powers they embed.

Notes

1 This was a flagship government policy introduced in 2001 to promote the 'golden age of business' through public-private sector participation to create wealth. Eventually, the private sector was to take over and manage it. The concept first sought to link farmers to existing mills, additionally establish new mills through strategic partners and encourage farmers' ownership through the concept of Corporate Village Enterprise (CoVE).

2 The Ghana COCOBOD is a state-owned regulatory institution mandated for the purchase and fixing of the buying price of all cocoa beans in Ghana. The institution oversees the production, processing, and marketing of cocoa in Ghana.

References

Agyemang-Duah, W., Kafui Gbedoho, E., Peprah, P., Arthur, F., Kweku Sobeng, A., Okyere, J., & Mengba Dokbila, J. (2018). Reducing poverty through fiscal decentralization in Ghana and beyond: A review. *Cogent Economics & Finance*, *6*(1), 1476035. https://doi.org/10.1080/23322039.2018.1476035

Ankamah, S. S. (2012). The politics of fiscal decentralization in Ghana: An overview of the fundamentals. *Public Administration Research*, *1*(1), 33. http://doi.org/10.5539/par.v1n1p33

Ankrah, D. A., Freeman, C. Y., & Afful, A. (2020). Gendered access to productive resources – evidence from small holder farmers in Awutu Senya West District of Ghana. *Scientific African*, *10*, e00604. https://doi.org/10.1016/j.sciaf.2020.e00604

Ayee, J. R. (2013). The political economy of the creation of districts in Ghana. *Journal of Asian and African Studies*, *48*(5), 623–645. https://doi.org/10.1177/0021909612464334

Bawole, J. N. (2017). Pro-poor decentralization in Ghana: Exploring the facilitators and the limitations. *Administrative Theory & Praxis*, *39*(2), 122–139. https://doi.org/10.1080/10841806.2017.1309801

Behrman, J., Meinzen-Dick, R., & Quisumbing, A. (2012). The gender implications of large-scale land deals. *Journal of Peasant Studies*, *39*(1), 49–79. https://doi.org/10.1080/03066150.2011.652621

Cabral, L. (2011). Decentralisation in Africa: Scope, motivations and impact on service delivery and poverty, FAC Working Paper 20, Brighton: Future Agricultures Consortium. http://www.future-agricultures.org/publications/research-and-analysis/working-papers

Conyers, D. (2007). Decentralisation and service delivery: Lessons from Sub-Saharan Africa. *IDS Bulletin*, *38*(1), 18–32. Retrieved from https://opendocs.ids.ac.uk/opendocs/bitstream/handle/20.500.12413/8340/IDSB_38_1_10.1111-j.1759-5436.2007.tb00334.x.pdf?sequence=1

Crawford, G. (2004). *Democratic decentralisation in Ghana: Issues and prospects.* Retrieved from Leeds, United Kingdom: https://citeseerx.ist.psu.edu/viewdoc/download?doi=10.1.1.127.303&rep=rep1&type=pdf

Crawford, G. (2008). Decentralization and the limits to poverty reduction: Findings from Ghana. *Oxford Development Studies*, *36*(2), 235–258. https://doi.org/10.1080/13600810701702002

Ghanadistricts. (2021). Retrieved from http://ghanadistricts.com/

Goel, R. K., Mazhar, U., Nelson, M. A., & Ram, R. (2017). Different forms of decentralization and their impact on government performance: Micro-level evidence from 113 countries. *Economic Modelling*, *62*, 171–183. https://doi.org/10.1016/j.econmod.2016.12.010

Inanga, E. L., & Osei-Wusu, D. (2004). Financial resource base of sub-national governments and fiscal decentralization in Ghana. *African Development Review*, *16*(1), 72–114. https://doi.org/10.1111/j.1467-8268.2004.00085.x

Jütting, J., Corsi, E., Kauffmann, C., McDonnell, I., Osterrieder, H., Pinaud, N., & Wegner, L. (2005). What makes decentralisation in developing countries pro-poor? *The European Journal of Development Research*, *17*(4), 626–648. https://doi.org/10.1080/09578810500367649

Lanz, K., Prügl, E., & Gerber, J.-D. (2020). The poverty of neoliberalized feminism: Gender equality in a 'best practice' large-scale land investment in Ghana. *The Journal of Peasant Studies*, *47*(3), 525–543. https://doi.org/10.1080/03066150.2019.1602525

Manor, J. (1995). Democratic decentralization in Africa and Asia. *IDS Bulletin*, *26*(2), 81–88.

Mockshell, J., & Birner, R. (2015). Donors and domestic policy makers: Two worlds in agricultural policy-making? *Food Policy, 55*, 1–14. https://doi.org/10.1016/j.foodpol.2015.05.004

Ndegwa, S. N. (2002). *Decentralization in Africa: A stocktaking survey*. World Bank, Africa Region.

Nyantakyi-Frimpong, H., & Bezner Kerr, R. (2017). Land grabbing, social differentiation, intensified migration and food security in northern Ghana. *The Journal of Peasant Studies, 44*(2), 421–444. https://doi.org/10.1080/03066150.2016.1228629

Prud'Homme, R. (1995). The dangers of decentralization. *The World Bank Research Observer, 10*(2), 201–220.

Prügl, E. (2015). Neoliberalising feminism. *New Political Economy, 20*(4), 614–631. https://doi.org/10.1080/13563467.2014.951614

Resnick, D. (2017). Democracy, decentralization, and district proliferation: The case of Ghana. *Political Geography, 59*, 47–60. https://doi.org/10.1016/j.polgeo.2017.02.011

Ribot, J. C., Agrawal, A., & Larson, A. M. (2006). Recentralizing while decentralizing: How national governments reappropriate forest resources. *World Development, 34*(11), 1864–1886.

Riedl, R. B., & Dickovick, J. (2010). *Comparative assessment of decentralization in Africa: Ghana desk study. Report prepared for the United States Agency for international development*. Retrieved from Washington, DC: http://www.usaid.gov/our_work/democracy_and_governance/technical_areas/governance/decentralization_in_africa.html

Riedl, R. B., & Dickovick, J. T. (2014). Party systems and decentralization in Africa. *Studies in Comparative International Development, 49*(3), 321–342. https://doi.org/10.1007/s12116-013-9144-9

Robinson, N. (2007). Citizens not subjects: US foreign relations law and the decentralization of foreign policy. *Akron Law Review, 40*, 647. Retrieved from https://heinonline.org/HOL/LandingPage?handle=hein.journals/aklr40&div=27&id=&page=

Smoke, P. (2003). Decentralisation and local governance in Africa-Preface. *Public Administration and Development: The International Journal of Management Research and Practice, 23*(1), 7–16. https://doi.org/10.1002/pad.255

Tsikata, D., & Yaro, J. A. (2014). When a good business model is not enough: land transactions and gendered livelihood prospects in rural Ghana. *Feminist Economics, 20*(1), 202–226. https://doi.org/10.1080/13545701.2013.866261

UNDP (2021). Environmentally sustainable production practices in cocoa landscapes (ESPII). https://www.gh.undp.org/content/ghana/en/home/operations/projects/environment_and_energy/ESPII.html. Accessed 07/10/2021.

Section III

Rights to food, land, and gender equality

Joanna Bourke Martignoni

Introduction

This section examines the interplay between agricultural and land commercialization and plural legal norms relating to gender equality, food, and land. In particular, it focuses on the ways in which the human right to adequate food, which is one of the core elements of the DEMETER research, is mediated through gendered laws, jurisprudence, and customs. The chapters highlight how international human rights law, national constitutions, legislation, and jurisprudence articulate with local customs, practices, and ideas of justice in connection with the allocation of ownership and usufruct rights over farming land and natural resources.

International human rights and development organizations, including the UN Committee on the Elimination of Discrimination Against Women (CEDAW), have frequently adopted discourses that argue that customary laws and practices are inherently antithetical to the realization of women's equal rights to food and land (Bourke Martignoni, 2018). As a result, these bodies have tended to privilege the 'formalization' of land titling systems alongside legislative anti-discrimination provisions as mechanisms to guarantee *de jure* gender equality in land inheritance, ownership, and use. Crucially, however, a wide array of 'informal' customs, practices, and adjudicatory mechanisms continue to exist in the same spaces as the statutory provisions and institutions that govern land use and, by extension, the right to food. Processes of food and land commercialization are causing these plural norms to be recast in multiple ways that vary over time and place and which have a diverse array of human rights impacts.

The two chapters on Cambodia in this section discuss the existence, on paper, of constitutional and legislative guarantees of formal gender equality in property ownership and inheritance, social protection, and an adequate standard of living. In his analysis of the Ghanaian constitution, Atupare notes that the right to non-discrimination on the grounds of sex is contained in the operative section of the document whereas the right to food has had to be inferred from judicial interpretations. In practice, however, the legal

DOI: 10.4324/9781003202004-11

protections accorded to human rights in Ghana have been more widely accepted than they are in Cambodia where there is no jurisprudence on the right to food or on most internationally recognized human rights, and the legal system is frequently used as a 'weapon' to further the interests of the economically powerful (Bourke Martignoni, 2021). In their discussion of land inheritance in Cambodia, Beban and Bourke Martignoni document the way in which the state is simultaneously present as the mediator of global capital and absent as a provider of social protection. The chapter by Bourke Martignoni and Joshi further highlights the linkages between the Cambodian government's facilitation of microcredit for women and the resulting land loss that can occur due to over-indebtedness. In the chapters on Ghana, Torvikey and Atupare also draw attention to the manner in which formal legislative guarantees of gender equality are often over-ridden by private economic interests.

All of the authors in this section demonstrate that the simple enactment of legislative guarantees of equal rights is insufficient to turn these into reality in settings where patriarchal power structures continue to determine access to land and natural resources. The chapter by Torvikey and Atupare on jurisdictional multiplicity in Ghana and the paper by Bourke Martignoni and Joshi on Cambodia attest to the existence of a 'justice façade' or a performance of equality through legislative provisions granting equal rights to own and inherit land without considering the multiple material ways in which women experience and resist dispossession in practice.

The section also examines how non-economic ontologies of land continue to guide family and community decision-making even within the statutory land titling regimes that serve to underpin land and agricultural commercialization. The role played by land as a form of informal social security in rural communities is discussed in the chapter by Beban and Bourke Martignoni and the authors draw attention to 'moral economies of care' that have not been overwritten by the privatization of formerly communal land or by the adoption of joint land titling and inheritance legislation. In their discussion of Ghana, Torvikey and Atupare examine local concepts of 'gender justice' and the ways in which these are constructed by customary decision-makers in cases involving intra-familial disputes over land rights. The ambiguities of customary land inheritance practices for women are illustrated in both of these chapters that show how matrilineal inheritance norms are being adapted to new situations of land scarcity with varying outcomes for gender equality and the right to food.

In each country, women are navigating plural normative systems in order to assert their rights to food and land. In their chapter, Bourke Martignoni and Joshi discuss the potential for the reconfiguration of gendered subjectivities through women's leadership of community protests against land grabbing in Cambodia. As Torvikey and Atupare show, in Ghana, women wishing to claim their rights to land or equal remuneration for agricultural labour are strategically 'forum shopping' in order to find a suitably receptive adjudicatory mechanism. In both settings, women's margin for manoeuvre is relatively constrained as a result of the workings of patriarchal patronage networks and legal institutions that tend to privilege the rights and interests of agribusinesses, local chiefs, and other powerful rural landholders. The chapters also document the way in which gender intersects with ethnicity, class, migration status, and age in determining who has access to justice in these settings of agricultural and land commercialization.

While international human rights law – with its emphasis on non-discrimination and the equal accessibility of food and land – provides a framework for the development of national legislation and policies, processes of normative appropriation and 'vernacularization' are dynamic and complex (Merry & Levitt, 2017). This section demonstrates the ambivalence of normative pluralism and the ways in which equality provisions in statutory laws are often disconnected from structural inequalities and therefore unable to act as an effective counterweight to patriarchal systems of power. In both Cambodia and Ghana, 'informal' norms and practices play a decisive role in influencing land governance and these customary norms are continually being adapted and remade in light of the new pressures on land induced by commercialization. The chapters in this section highlight the need for greater inter-disciplinary engagement with material experiences of normative pluralism and the multiple ways in which the right to food might be realized or violated in different contexts.

References

Bourke Martignoni, J. (2018). Engendering the right to food? International human rights law, food security and the rural woman, *Transnational Legal Theory*, 9:3–4, 400–414. https://doi.org/10.1080/20414005.2018.1568789

Bourke Martignoni, J. (2021). A feminist methodology for implementing the right to food in agrarian communities: Reflections from Cambodia and Ghana, *The Journal of Peasant Studies*. https://doi.org/10.1080/03066150.2021.1928642

Merry, S. E., & Levitt, P. (2017). The vernacularization of women's human rights. In *Human Rights Futures*. Cambridge University Press, pp. 213–236. https://doi.org/10.1017/9781108147767.009

8 Feminist legal geographies of land titling, indebtedness, and resistance in rural Cambodia

Joanna Bourke Martignoni and Saba Joshi

Following almost 30 years of armed conflict, the Cambodian peace agreements of the early 1990s ushered in several decades of intense policy and legislative activity related to land tenure reform, which is viewed as integral to the success of democratic governance and economic development in the country. The emphasis within cadastral programmes on joint land titles, issued in the names of both spouses, is based on the widely held assumption that equal ownership and control over land will lead to women's empowerment in both the 'public' and 'private' spheres along with a host of other desirable development outcomes, such as improved agricultural productivity, child nutrition, and educational attainment. This approach is supported by a range of international development actors active in the Cambodian land sector.

While women's equal rights to use, own, inherit, and control land are seen as core components of development programmes, these initiatives tend to conceive of land as an individually held economic asset without attending to relations of power that are embedded within the norms and practices of attributing land rights in rural communities (Li, 2014; Razavi, 2003; Schoenberger and Beban, 2020). The context within which formal, gender-equal rights to land have been accorded in Cambodia is one of highly unequal neo-liberal economic development (Beban, 2021; Kent, 2016). As Katherine Brickell notes in her study of the relationship between domestic violence and forced eviction in Cambodia, the existence of a 'justice façade' or a performance of equality in the formal legal provisions on land inheritance and ownership is at odds with the multiple material ways in which women experience and resist dispossession in practice (Brickell, 2020). In this dynamic setting, social class, age, and ethnicity intricately intertwine with gender identities to produce distinct experiences of oppression and domination (Park & Maffii, 2017).

The discussion in this chapter seeks to draw out the ways in which gender matters in the Cambodian land sector. By adopting a feminist legal geography framing, it also critically analyses the manner in which the roles and agency of women and men in rural spaces are reflected in laws and debates over land and resource rights (Bourke Martignoni, 2018; Brickell, 2020; Federici, 2019). First, we outline the legal and policy frameworks developed

DOI: 10.4324/9781003202004-12

by the Cambodian government and inter-governmental organizations, as well as the unwritten norms and the violence that govern the allocation of rights to own and use land and natural resources in Cambodia. We argue that Cambodia's current land governance regime, which theoretically enables formal land ownership for women through joint titling and subsequent access to micro credit, simultaneously emerges as a site of dispossession with distinctly gendered consequences (Brickell, 2020). Smallholder farmers are provided with private land title in a highly competitive environment that privileges investor-friendly large-scale land acquisitions through the economic land concession (ELC) policy introduced under the 2001 Land Law, the unequal effects of which have not been offset through the creation of social land concessions (SLCs). The failure of these policies to ensure equitable outcomes – coupled with the fact that both private and communal land titling[1] have not been implemented consistently across the country – exposes the gendered geographies of the commercialization of both land markets and agricultural production.

In the second part of the chapter, we consider the exponential growth of microfinance institutions and the way that debt and the accompanying potential for land loss through repossession are imbricated within the neoliberal promotion of private land titling. Arguments made in this section show how the agrarian transition in Cambodia is refracted through gendered power relations that determine access to land and capital. These relationships are further mediated through structures of power that are contingent upon attributes such as indigeneity, socio-economic class, and age as well as local histories of land use and agricultural production.

In the final section of the chapter, we explore the ways in which land dispossession – produced at the intersection of state and market forces and frequently legitimized through the legal system – becomes a driver of public resistance, articulated through distinctly gendered political subjectivities. Examining women's participation in movements contesting land dispossession, we show how gender constructs – meanings, roles, and ideals associated with notions of femininity and masculinity – are recast in violent struggles over rights to land and natural resources (Brickell, 2020). Women's resistance to the expropriation of their homes, farms, and communally managed lands thus reveals how agrarian transformation is deeply embedded within gendered social relations and the ways in which these intimate, localized conflicts are connected to global supply chains that promote the liberalization of land and agricultural markets.

Our arguments are based on a feminist legal geographies approach that emphasizes the complex materialities of power relations within the 'law-space nexus' of land governance in rural Cambodia (Cuomo & Brickell, 2019, p. 1047). This framework allows us to make visible the temporal and spatial linkages between international and national legal and policy frameworks and their articulation with local practices of land titling, dispossession, and resistance (Brickell, 2020). Along with an analysis of relevant gender equality provisions in land laws and policies, we draw upon empirical findings generated from semi-structured interviews carried out by DEMETER researchers over multiple field visits in the three Cambodian provinces of Kampong Thom, Ratanakiri, and Kratié between 2015 and 2020. Research participants included community leaders (commune officials, village chiefs, and deputies) and households (both women and men). Our research was further informed by interviews with national civil society actors – based in Phnom Penh as well as in the three provinces – engaged in the fields of land and natural resource governance, gender and development, and human rights. These interviews were digitally recorded, transcribed, and coded using the qualitative data analysis software, Nvivo 12. Our coding protocol included a mix of inductive and deductive methods, and the key themes discussed in this chapter were drawn out through multiple rounds of qualitative coding.

The role of the state in shaping gender (in)equality in written and unwritten land laws and policies

Private titles, joint land tenure, and large-scale land acquisitions

The gender dimensions of processes of land-use change in Cambodia are difficult to quantify, given the lack of accessible and reliable data concerning land ownership and use in the country as a whole (Kieran et al., 2015). In a comparative survey of gender inequality in a number of Asian countries, Kieran et al. conclude that land ownership is 'very common' among both women and men in Cambodia, that men are only slightly more likely than women to be landowners, and that a 'higher percentage of women than men own some land solely' (2015, p. 127). The demographic survey data on which the authors base their findings has been used by government and development institutions to underline the success of the joint titling approach that was introduced in the 2001 Land Law. In 2014, the Cambodian Ministry of Land Management, Urban Planning, and Construction published statistics that showed that spouses jointly held 63% of the land titles issued in the country (Joshi, 2018). Other data indicate that in 2017, female-headed households owned 15% of Cambodia's agricultural land (Cambodian government, 2017, pp. 21–22).

These findings are tempered by the fact that, in 2017, almost 60% of rural households reported that they had land holdings of less than 1 ha (Cambodian government, 2017, p. 23). In DEMETER survey data from the three Provinces (Kratié, Kampong Thom, and Ratanakiri) that were sampled in 2016, there was significant land loss amongst smallholders as a result of the extension of large Economic Land Concessions (ELCs) and this process of agrarian change affected 74% of the land that was held by those farmers with the smallest amount of land (less than 2 ha), whereas those with land holdings over 7 ha reported that they had still lost 36% of their land over the preceding five years. Importantly, female-headed households are much more likely to have smaller land holdings and are, therefore, disproportionately vulnerable to land loss (Gironde et al., 2021). Declining average farm sizes and the consolidation of land holdings into large agribusiness operations mean that most smallholders can no longer rely on agricultural production as their main source of income or food (Gironde et al., 2021).

In spite of the emphasis on national policies, and legislation on the importance of land mapping, titling, and registration, private land titles have not reached the majority of Cambodia's rural population, particularly in areas that are historically tenure-insecure and that have been targeted for land concessions (Beban, 2021; Kent, 2016; Peou & Young, 2019). A 74-year-old widow in our study noted the difficulties associated with transmitting farming land in such an uncertain climate and described the common experience of being dependent on the local government authorities for title:

> I really want to have land title. I want to have it so that it is easy to take care of the land. For the land that I gave to my children, I also want to have land title for their land, so it is easy for them to take care of their land. But the authority has not come to provide land title to us.
>
> (74-year-old woman, Kampong Thom, March 2020)

Mapping of land titling programmes and land concessions indicates that systematic titling was avoided in areas where a majority of ELCs were granted (Diepart & Schoenberger, 2017; Dwyer, 2015; Park & Maffii, 2017). Thus, people who farmed on land that was likely to overlap with ELCs did not receive land titles and also lost access to the common forest and grazing lands and fisheries that have historically acted as an important source of non-farm income and food security (Joshi, 2018).

In 2012, the Cambodian government implemented a rapid land titling regulation (Order 01) aimed at quelling land conflicts between ELCs and rural citizens, enacting what it called the 'leopard skin' (*sbaich klar*) policy, where small farms would be carved out of large tracts of concession territories or state-held communal lands (Beban, 2021). Our interviews in villages in Kratié and Ratanakiri revealed that gaining land titles for farming land on the so-called 'leopard's spots', surrounded by large agro-businesses, did not secure them against dispossession. A village chief in Ratanakiri explained how, despite possessing land titles, households situated close to ELCs often lost their lands:

> The company sometimes encircles people's land and threatens to move them out of that land. In return, they offer villagers some small money, around USD 200 per hectare, as compensation. They tell them that they will lose their land in any case so it's better to accept the compensation.
>
> (Male village chief, Ratanakiri, 2016. Also cited in Joshi 2020, p. 5)

Any findings about the role of land mapping and titling initiatives with respect to advancing gender equality therefore need to be read against the politics of land concessions wherein the mere existence of mapping and registration processes does not automatically lead to security of tenure for either women or men. The importance of class status, political connections, and geographical location in determining access to land titles poses additional challenges for women, who are typically less embedded in local patronage structures (Maffii, 2009; Kusakabe et al, 1995; Kusakabe, 2015). A 30-year-old woman, recently separated from her partner, spoke about the bribes she may have to pay to local officials to secure title for her land:

> I want to get a land title, but I know how much it costs to ask the village head. If officials ask me for extra money, I wouldn't be able to pay it.
>
> (30-year-old woman, Ratanakiri, 2016)

The gender of joint land registration and titling

The 1993 Cambodian Constitution contains guarantees of equal land and property rights for all Khmer citizens as well as a prohibition on discrimination against women in line with the country's ratification of international human rights instruments, including the United Nations Convention on the Elimination of All Forms of Discrimination Against Women (CEDAW). Under the 1989 Cambodian Law on Marriage and the Family, spouses have equal rights to property acquired during marriage and to individually own property possessed prior to marriage or received as a gift or inheritance. Joint property may not be disposed of without the consent of both spouses. The 2011 Civil Code further strengthens the legal framework for formal gender equality in rights over immovable property in relation to marriage, divorce, inheritance, and intra-familial succession.

Despite the evidence mentioned above that systematic land registration schemes have not increased tenure security for most people in rural areas, the commonly held view amongst development actors in Cambodia is that joint land titling, whereby property is registered in the names of both spouses, promotes women's empowerment both in the 'public' domains of markets and political institutions and in the 'private' sphere of the family (Thiel, 2010). Several of the female participants in our research also expressed the opinion that their power within the household was deeply connected to the formal acknowledgement of shared land ownership and usufruct rights.

INTERVIEWER: It seems your husband is the main decision maker in your family?

RESPONDENT: Yes, because the land is under his name.

(28-year-old woman, Kratié, 2017)

Women's land tenure, however, has multiple meanings although joint title is often regarded as a proxy for the full exercise of ownership, access, usufruct, and transfer rights over land. There are many structural limitations that interact with contextual factors to influence the extent to which spouses with joint title are equally able to exercise the 'bundles of rights' that attach to tenure (Doss & Meinzen-Dick, 2020). In most of the households that were surveyed in the DEMETER research, married couples with joint ownership of land reported that although women often hold family money for safekeeping and are responsible for decisions about food, health care, education, and household expenses, it was the male partner who was most likely to take decisions with respect to buying and selling land as well as which crops should be planted on it. Men's ability to take unilateral decisions on land inherited by their partners reveals the unequal power relations in household decision-making over shared assets:

> I had 2 hectares of land that I inherited from my parents. My ex-husband was a gambler. He lost a lot of money and sold my land. I divorced him because he was always in debt, and we were always fighting.
>
> (30-year-old Charay woman, Ratanakiri, 2016)

While the woman quoted above was able to obtain an official divorce certificate, few men and women in rural Cambodia formally register marriages or divorces with local authorities. As a result, when couples separate, joint titles can impose an additional complication as changing legal proof of ownership is often too costly and difficult for poor households (Joshi, 2018; Kusakabe et al., 1995). While local authorities tend to acknowledge *de facto* marriages, they typically do not do the same for separations and our research documents many cases of village and commune officials pushing couples to reconcile, including cases where there was a history of domestic violence or an otherwise seemingly irretrievable breakdown in the relationship (Joshi, 2018). In this way, women are often pressured by authorities to maintain their former partners' name on the joint land title documents, even if they do not wish to do so, to preserve at least the outward appearance of a harmonious settlement (Baaz et al., 2017). The Cambodian situation thus reveals the ways in which linking land ownership to marital status through joint titling of family land does not inexorably lead to women's emancipation and may, in certain circumstances, actually contribute to their dispossession.

Under customary law, Cambodian women have the right to unilaterally inherit property and to withhold it after divorce. However, these practices are being supplanted by formal statutory law and the presumption of an equal division of marital property in the event of dissolution of the partnership. The Civil Code, adopted in 2007, provides that upon divorce, marital property and assets shall be divided equally between spouses (Van der Keur, 2014). In recognition of this, several of the respondents in our qualitative interviews stated that their mothers were reluctant to transfer land title to their married daughters, not wishing it to go to the husband in the event of divorce. One of the indigenous women in our survey area reflected on this issue:

> My mother just shares the land so that I can work on it. She dare not inherit land to me for land title because I got married to a Khmer man. She is afraid that my husband would leave and sell the land. So, I just keep working on her land and get harvest for sale.
>
> (25 year-old woman, Ratanakiri, 2020)

Another crucial disadvantage of joint titling for women's empowerment is that many women remain in abusive relationships due to fear of losing their share of land in the event that they decide to leave. Several women in our research recounted feeling 'trapped' as a result of the legal, financial, and social difficulties that extricating themselves from situations of joint ownership would entail . Similar findings have been reported by researchers working in other areas of the country (Baaz et al., 2017).

Land redistribution through SLCs

The 2001 Land Law established a dual track system of ELCs, designed to promote agricultural investment and to create employment opportunities, and SLCs that were supposed to enable landless and poor people and groups including veterans and persons with disabilities to access housing and farming land (Neef et al., 2013). Women were one of the target groups identified as potential beneficiaries of the SLC programme, which was heavily supported by the World Bank with a view to providing the poor with land as an economic and food-producing asset that could be further leveraged in order to access credit to expand their incomes (Lamb et al., 2017; Neef et al., 2013; Thiel, 2010). An additional rationale for the allocation of SLCs, provided for in the 2007 Sub-Decree on Social Land Concessions, is that they may function as resettlement sites for populations displaced by ELCs and, therefore, contribute a pool of labourers linked to the economic concessions to promote the development of agro-industries. In reality, the role played by ELCs in the creation of rural employment in local communities, particularly for women, has been relatively insignificant and the idea that displacing communities could be made more palatable through the offer of land and housing in peripheral areas with few infrastructures remains contested (Kusakabe, 2015; Diepart & Schoenberger, 2017; Joshi, 2020; OHCHR, 2018).

In 2018, the UN Office of the High Commissioner for Human Rights (OHCHR) in Cambodia conducted an assessment of six land allocation projects established under the National Social Land Concession scheme. They surveyed 640 households and found that 28% of respondents declared that the head of the household was female, with 82% of those interviewed stating that they were married or in a domestic relationship, whilst 13% were widowed or divorced (OHCHR, 2018, p. 17). On the same page, the OHCHR notes:

> Gender discrimination in terms of land allocation and distribution does not appear, *prima facie*, to be an issue, although monitoring and data

collection appears to be inconsistent across the six SLC sites. Gender discrimination does appear to be a persistent problem in terms of access to education.

The OHCHR assessment found that while average income had increased for households living in SLCs, there were serious problems with respect to food insecurity, lack of water and sanitation as well as with the soil fertility of the land allocated for agricultural production. In addition, the issue of conflicts between SLC beneficiaries and local communities was highlighted – in particular, the clearing of communal forest land for the concession – as well as the lack of responsiveness of authorities at the village and commune levels of governance in connection with complaints about conditions in the sites (OHCHR, 2018).

One of the communities in the DEMETER study area includes an SLC and our findings about the impact of the allocation of land on the existing villagers as well as the beneficiaries of the project are very similar to those contained in the OHCHR assessment. Many of the people interviewed in 2016, and again in 2020, appeared unclear about who the beneficiaries of the SLC are: some stated that it was persons with disabilities, others, that it was veterans, widows, or people identified by social security services as poor and vulnerable. A number of those we spoke to pointed out the large numbers of empty dwellings and the lack of active farming going on in the community despite the stipulations in the 2007 Sub-Decree on Social Land Concessions that provide that the land allocated for SLCs must be resided upon for at least six months per year and cultivated within 12 months of it being granted. During our round of qualitative interviews in 2016, a 45-year-old widow recounted the way in which she had originally been allocated land in the SLC, but the authorities decided to convert the area into an ELC and effectively evicted the population by depriving them of food, health care, and freedom of movement until they agreed to be resettled in the village. Following that experience, she had spent many years embroiled in negotiations concerning usufruct rights over her farming land:

A: I currently own two plots of residential land. One plot I got from the committee when they moved me out of the ELC area. And one more plot I bought from a person whose name was on the list to get land as compensation from the committee when they were moved out of the ELC.

Q: How about farmland?

A: I have two plots of *chamka* (farm) land. One plot is 1 ha and one is 1.5 ha. 1 ha of *chamka* land I received from the social land concession and 1.5 ha I bought from a neighbour for 700USD. Both plots I cannot access for farming even though I got the land receipt from the village chief in 2013. I have the land receipt but have never been able to use the land for farming.

(45-year-old woman, Kampong Thom, 2016)

The contribution of SLCs to gender equality in land rights needs, therefore, to be weighed against the way in which these schemes have been implemented in practice. As mentioned in the introduction to this section, the political economy of rural Cambodia is such that statutory land laws do not automatically prevail over other forms of normativity grounded in patronage relationships, administrative practices, or even traditional usufruct rights (Adler & So, 2012). The prospects of land redistribution for women through SLCs are, therefore, relatively limited and demonstrate the ways in which poverty, ethnicity, and gender intersect in particular settings to produce specific experiences of inequality and exclusion (Lamb et al., 2017; Neef et al., 2013; Park & Maffii, 2017). The fact that most local political authorities, including those responsible for the governance of various SLCs, are male-dominated, means that these structures do not necessarily provide the opportunities for women's empowerment or gender equality in the public sphere that have often been anticipated as a desirable side effect of the establishment of government-led land redistribution schemes (Lamb et al., 2017).

The 'crisis ordinary': gendered geographies of indebtedness and land loss

A further threat to security of land tenure for smallholder farmers and rural labourers has emerged in Cambodia in recent years in the form of commercial microfinance actors that provide loans against the land titles that they hold as collateral. Several non-governmental organizations and researchers have sounded the alarm over the unethical lending practices that characterize the relationship between financial institutions and the Cambodian land sector (Bateman, 2020; Green & Bylander, 2021). To date, however, this form of dispossession has generally been regarded as a 'crisis ordinary' rather than a serious or urgent form of structural violence that demands an immediate political response (Brickell, 2020). The gender dimensions of over-indebtedness and land repossession have yet to be comprehensively evaluated. However, many of our interviewees – both women and men – reported high levels of stress connected to the pressure to repay loans, along with their fear that default will lead to land loss and the transmission of debt to other family members.

Within the new neo-liberal landscape, where rural people are encouraged to leverage their assets and become entrepreneurs, the necessity to use land as collateral for loans pushes smallholder farmers to seek formal title for their land from local authorities. As a 51-year-old married woman with several adult children noted during an interview with the DEMETER research team in 2017:

> Currently many people get loans from the microfinance institutions (MFIs) or banks. Before mostly people got loans from private money-lenders but since then a lot of MFIs and banks come into the village and

> many people change to access loans from them. As I said, next year I will ask for land title to make it easier to get a loan.
>
> (Kratié, 2017)

An overwhelming number of respondents in our surveys reported juggling multiple loans from commercial financial institutions and from informal money lenders to manage ever-increasing levels of debt (DEMETER, 2021). Most of the households we interviewed stated that both partners had jointly made the decision to take out loans for farming inputs, new small businesses such as grocery stores, repayments on existing loans, medical costs, or everyday items such as food and educational expenses. Several of the loan officers from microfinance institutions (MFIs) whom we spoke to claimed that they were legally required to ensure that both of the land title holders gave their consent to the loan and that it was crucial that they were joint parties to loan agreements. Nevertheless, in practice, our research shows that it is women who often bear the day-to-day emotional load of navigating repayments.[2] A married Khmer woman in her mid-30s recounted that she and her husband had used their jointly owned farmland as collateral to borrow USD25,000 from a MFI for the purposes of building a house and starting a grocery and money transfer business:

> Q: The loan is under your name too. If there are any problems in your family, what will you do?
>
> A: I think if that happens, I will sell land to pay back the loan. In fact, the loan was under my husband's and my name from the beginning because we are both on the land title, but I am the one whose name is on the form that says that I will repay the money to the bank.
>
> (Kampong Thom, 2020)

This situation accords with the social norm in many Khmer communities that it is women who hold responsibility for family financial management. While this clear-cut division of household labour is not as prevalent in indigenous communities, one of our male respondents noted that, in the Charay and Tampuon ethnic groups in upland provinces in the country, seeking assistance from informal family networks to repay loans was also regarded as falling to women:

> Normally, the wife will look for a solution like borrowing money from her siblings or parents. It is rare for the husband to borrow money from his parents to repay loans because in Charay and Tampuon culture, parents help their daughters more than their sons after marriage.
>
> (28-year-old indigenous man, Ratanakiri, 2019)

Other female respondents discussed the way in which using jointly titled land as collateral meant that they were trapped by the looming threat of land repossession in a seemingly never-ending cycle of repayments. A 65-year-old woman from Kampong Thom whose husband had decided to become a monk and retire from their farm vividly discussed her current responsibility for maintaining the repayment schedule on her own, despite the fact that her financial situation was now radically different.

> When we took the loan, they asked for the land title and how many places I had loans from. I had never had a loan before. They asked me to give my thumb print at the village. I gave my thumb print with my husband, and I had a witness and the witness also gave thumb prints on their documents.
>
> Q. Does this mean that you still have to pay back the loan even though one of you is missing?
>
> A. Yes, I have to pay. For example, I have to pay on the third of every month, but I cannot pay. So they will fine me because I am late.
>
> (Kampong Thom, 2020)

The impact of microfinance debt on the accessibility of the rights to food and health is also clearly apparent in our research. In the 2020 round of interviews, a woman in her 40s from Kampong Thom was visibly upset about the way in which indebtedness was affecting her family's food security and the fact that they were trapped in a vicious cycle whereby selling land to repay loans would lead to their having even less access to income from cash crops or to food grown through subsistence farming:

> We don't dare eat delicious food. We need to save money to repay the debt. Before we had a loan, we could eat 1kg of fish in two days. Now we have to make the fish last for one week. We think that if we sell the land in order to repay the debt, we will not have land to do farming and we will have even less food to eat.
>
> (Kampong Thom, 2020)

The interviews also reveal the ways in which the gender division of reproductive and productive labour within families with respect to loan repayments reflects the political economy of rural labour markets. Many of the people surveyed noted that it is husbands and sons who bear the primary responsibility for earning income to service familial debt as women and older people are generally not able to find employment that would enable them to also care for children and other dependent family members (Joshi, 2020). There are a variety of experiences here, however, with some women reflecting on the fact that they can no longer 'stay at home, doing the domestic work' because they must contribute income for loan repayments or risk losing the family's land (36-year-old Khmer woman, Kampong Thom, January 2016). On the other hand, a 30-year-old married mother of two commented on the way in which indebtedness had further cemented the traditional gender division of household reproductive labour in her family:

Q. You now have more loans than before. So, does your family work harder than before or not?
A. Yes, we try harder and we do not have much time to relax.
Q. When you had smaller loans, did you have to work this hard?
A. Before, we had time to stay at home. But when we took more loans we cannot stay with family, we need to earn money. For me, I stay at home to take care of children. My husband and I work harder than before. I pity my husband, but we need to do it for our family. We took the loan to buy equipment so when we pay off the MFI that equipment will become our property.
(Kampong Thom, 2020)

These gendered experiences of indebtedness tied to the potential for land loss are also mediated through characteristics such as age, ethnicity, and social class. Several of our respondents recounted stories of mixed marriages in which indigenous women had their inherited land mortgaged by their non-indigenous husbands without their consent. To overcome the potential loss of family land in this way, some indigenous women noted that their mothers were no longer passing their land down to them but were, instead, allowing their daughters to work on the land and collect money from the harvest. However, this strategy was not always successful in preventing dispossession through indebtedness. In one of our interviews in Ratanakiri, a respondent recounted a story about a couple where the Khmer husband had fraudulently used his indigenous mother-in-law's land certificate to borrow money from a cassava broker against the value of her crop and land:

> The husband (Khmer man) went alone to the broker while his wife went to the farm. He showed the land certificate which belonged to his mother-in-law to the cassava buyer… So, finally his mother-in-law had to pay back the loan instead. She paid him back by selling 1ha of land to get 1,500 USD.
> (42-year-old indigenous man, Ratanakiri, 2019)

As discussed in Chapter 10 in this collection, the absence of social protection mechanisms such as state-provided pensions also has an impact on the ability of many rural people – particularly women, who tend to be excluded from the formal labour market and its social security entitlements – to continue to service debt (Hiilamo et al., 2020). Many older people in our research, such as this woman in her late 50s with three adult children, highlighted their anxiety about intra-familial debt transmission:

Q. Do you still have other loans?
A. No, I don't want to borrow anymore. We are old now and we can save some money to meet our needs. If we need more money, we can sell

> labour instead. We are afraid that if we borrowed money and accidentally died, our children would take over our debts. If they cannot earn money to pay back the loan, they would lose their land.
>
> (Kampong Thom, 2019)

The expansion of microfinance throughout the Cambodian countryside is remaking the landscape and increasing the precariousness of already insecure tenure for many smallholder farmers (Green & Bylander, 2021). While the gendered contours of land use change arising from indebtedness still need to be studied, this context of intensive commercial pressure on land has led some women to take on new roles to resist multiple forms of dispossession.

Standing one's ground: women's resistance against land dispossession and commercialized agriculture

Over the past fifteen years, women have been increasingly involved in social movements that seek to contest land and resource grabbing throughout Cambodia. The presence of women as leaders of resistance movements in high profile land conflicts, including the notorious Boeung Kak lake development in Phnom Penh, which resulted in six women serving lengthy prison sentences, has been the focus of much feminist scholarship on Cambodia (Brickell, 2014; Hennings, 2019; Joshi, 2020a; Lamb et al., 2017; Park, 2019; Park & Maffii, 2017). Such protests draw attention to the ways in which constructions of femininities and masculinities and affective relationships with land influence the forms taken by social dissent as well as the ways in which gender stereotypes are strategically deployed by the actors involved (Hennings, 2019; Joshi, 2020a). Resistance to land-use change also acts as a site for the renegotiation of prevailing gender norms and points to the deep imbrication of gender hierarchies within political processes of claim-making (Beban, 2021; Brickell, 2014, 2020; Lamb et al., 2017; LICADHO, 2014; Park, 2019).

Several of the respondents in the first round of DEMETER qualitative interviews in 2016 spoke powerfully about their experiences of engaging in resistance to violent, state-supported, corporate land grabbing, and of the gender dimensions of these protests. There was a widely held assertion that women were more adept at negotiating settlements than men and that part of their motivation to take part in protests was the existence of gender norms that accord women with the primary responsibility to maintain 'land for farming to feed our families' (25-year-old woman, Kratié, 2017). In one case, a 48-year-old indigenous woman recounted the way in which her community had sought to defend their land from company bulldozers:

Q. Did anyone die or get injured when they were shooting?
A. They shot but the people ran away to escape the gun fire so no one died. After they stopped shooting, we went back to the places again. We

decided to struggle until we die. When they used an excavator to dig land for a canal, about 20 of us walked towards the excavator at the same time. Then they stopped digging but first they bulldozed all our crops and made a canal surrounding our lands.

Q. When you had the protest, was it women or men standing in the front line against the companies?

A. It was women and even young girls in the front line. Men dared not stand in front, they were afraid that the company would respond more violently if they did. If the women and girls in the front line were treated with violence, then the men behind would help. But finally, we still could not win.

(Kratié, 2016)

The mobilization of gendered stereotypes about women – mothers in particular – being naturally more inclined towards the promotion of peace and non-violence is one that has been a common feature of land conflicts throughout Cambodia (Hennings, 2019; Joshi 2020a; Lamb et al., 2017; Park, 2019). There is a diversity of viewpoints about the meaning of women acting as protagonists in anti-land grabbing movements. Some authors suggest that women's bodies are being 'instrumentalized' by non-governmental organizations and that pushing them as 'human shields' into the front lines in land protests, rather than leading to increased empowerment, actually exposes them to considerable physical, emotional, and financial risk that has not been rewarded with a reconfiguration of gendered social hierarchies (Hennings, 2019). Others, like Lilja, argue that while women's resistance is 'parasitic on traditional discourses', women land activists are disrupting gendered stereotypes through these repeated assertions of their rights (Lilja 2016, p. 687). In the same vein, it has been observed that groups of indigenous women in upland Cambodia are actively exercising their agency to claim land rights even in the face of family and community opposition and in the absence of external support (Park & Maffii, 2017; FAO, 2019).

At the same time, our research made visible important variations in modes and forms of women's protests across regions, each marked by local histories of dispossession, agrarian change, and gender norms. Khmer women based in Phnom Penh and lowland provinces, due to their connections with civil society networks and higher levels of education, were more visible as public protestors and their protest activities received coverage in national and international media (see, for example, CCHR, 2015; Wight, 2015). In our study areas, particularly in Ratanakiri, there were few women and men from indigenous communities who had travelled outside their village to participate in or organize protest movements. A Tampuon woman who heads a civil society organization based in Ratanakiri argued that while indigenous women play an important role in anti-dispossession struggles, they engage differently from their Khmer counterparts in other regions.

> Indigenous women's protest is very different from Khmer women. Our traditions are very restricting for women. Single women doing something without elder's permission is not allowed. The community encourages women to work in groups and no one becomes a 'leader' but rather women work as a collective.
>
> (36-year-old woman, Ratanakiri, 2019)

Our interviews also reveal the many ways in which women engage in less visible forms of everyday resistance to land dispossession and to agricultural commercialization. These include women acting as organizers by collecting thumb print signatures from community members affected by land grabbing and using these to successfully petition political authorities in provincial centres for compensation and the return of land from ELCs (Kampong Thom, 2016). In Kratié, a number of families – including several headed by women – discussed their resistance-by-occupation in which they camped on their farming land for over four years to prevent their fields from being taken by the ELC. Some of our respondents who were not personally affected by land grabbing also talked about providing financial and moral support to people in their communities who had lost land out of a feeling of solidarity. In other areas and at other times, women mentioned their practices of exchanging labour and participating in the preservation of communal forest and grazing lands as strategies to ensure the continuation of non-commercial circular economies (Female focus group discussions, Kratié, 2017; Ratanakiri, 2019).

Conclusions: the gendered legal geographies of land rights

The Cambodian case provides an illustration of the gendered legal geographies of land ownership and user rights in rural communities. In practice, legislative and policy measures designed to promote gender-equal land rights and employment opportunities in rural areas through presumptive joint titling, ELCs, and SLCs have not been able to redress the systemic inequalities generated by land and agricultural commercialization. Our research documents the ways in which gender intersects with ethnicity, social class, age, and location to produce specific forms of legally sanctioned dispossession.

In connection with women's access to land through state-supported redistributive schemes, Cambodian government agencies and development organizations highlight statistics that suggest women have been the primary beneficiaries of cadastral reform through joint titling and the allocation of SLCs since the early 2000s (Cambodian government, 2017; Thiel, 2010). In reality, these processes of formal tenure reform have failed to produce a sustainable system of land registration and titling and have arguably underpinned massive deforestation and privatization of common pool resources in the country, along with 'public' and 'private' violence and land loss as a result of over-indebtedness (Beban, 2021).

A feminist approach to the legal geographies of land use directs us to look at the ways in which the Cambodian government, international development institutions, and global markets are shaping both the 'public' space of land and agricultural investment laws and policies, as well as the 'private' space of intra-familial land ownership, and the household division of reproductive and productive labour (Brickell 2020; Cuomo & Brickell, 2019; Federici, 2019; FAO, 2019)). Feminist analysis also provides a counterweight to the dominant neo-liberal paradigm of land as a neutral, economic asset by attending to the ways in which land use is embedded within social relations of power at different geographic scales (Cuomo & Brickell, 2019; Schoenberger & Beban, 2020; Kelkar & Krishnaraj, 2020). A feminist exploration of material experiences of possession, dispossession, and resistance might take us beyond the existing market-centric approaches to gender equality and land governance in Cambodia towards a more equitable future.

Notes

1 Cambodia's 2001 Land Law recognizes indigenous peoples' rights to communal titling and offers protection against the acquisition of indigenous communal land by the state and individuals outside these communities. However, until 2021, only 33 applications for communal land titling had been granted by the Cambodian government (Keeton-Olsen, 2021), while many of the provisions for interim protection of indigenous lands have failed to be implemented.

2 See Chapter 1, 'From food-crop to food-shop: Agricultural commercialization and food security in Cambodia' in this collection.

References

Baaz, M., Lilja, M. and Östlund, A., 2017. Legal pluralism, gendered discourses, and hybridity in land-titling practices in Cambodia. *Journal of Law and Society*, *44*(2), pp. 200–227.

Bateman, M., 2020, March. Land titling improves access to microcredit in Cambodia: Be careful what you wish for. In *Presentation at the 2020 World Bank Conference on Land and Poverty*, Washington, DC.

Beban, A., 2021. *Unwritten rule: State-making through land Reform in Cambodia*. Cornell.

Bourke Martignoni, J., 2018. Engendering the right to food? International human rights law, food security and the rural woman. *Transnational Legal Theory*, *9*(3–4), pp. 400–414.

Brickell, K., 2014. 'The whole world is watching': Intimate geopolitics of forced eviction and women's activism in Cambodia. *Annals of the Association of American Geographers*, *104*(6), pp. 1256–1272.

Brickell, K., 2020. *Home SOS. Gender, violence and survival in crisis ordinary Cambodia*. Wiley.

Cambodian government, 2017. *Cambodia: Socio-Economic Survey 2017*. National Institute of Statistics, Ministry of Planning, Phnom Penh.

CCHR. 2015. 'Women on the Front-Line: The Situation of Human Rights Defenders in Cambodia in 2015.' Briefing Note. Phnom Penh, Cambodia: Cambodian Center for Human Rights (CCHR).

Cuomo, D. and Brickell, K., 2019, Aug. Feminist legal geographies. *Environment and Planning A: Economy and Space*, *51*(5), pp. 1043–1049.

Diepart, J.C. and Schoenberger, L., 2017. Governing profits, extending state power and enclosing resources from the colonial era to the present. In Brickell, K., and Springer, S. (eds.) *The handbook of contemporary Cambodia*. Routledge, pp. 157–168.

Doss, C. and Meinzen-Dick, R., 2020. Land tenure security for women: A conceptual framework. *Land Use Policy*, *99*, 105080.

Dwyer, M.B., 2015. The formalization fix. Land titling, land concessions and the politics of spatial transparency in Cambodia. *The Journal of Peasant Studies*, *42*(5), pp. 903–928.

FAO, 2019. *Women's land rights and agrarian change: Evidence from indigenous communities in Cambodia*. Phnom Penh, FAO.

Federici, S., 2019. *Re-enchanting the world: Feminism and the politics of the commons*. PM Press.

Gironde, C., Reysoo, F. and Torrico Ramirez, A., 2021. No cash, no food: Gendered reorganization of livelihoods and food security in Cambodia. *Journal of Peasant Studies*, *48*(7), pp. 1–22.

Green, W.N. and Bylander, M., 2021. The exclusionary power of microfinance: Over-indebtedness and land dispossession in Cambodia. *Sociology of Development*, 7(2), pp. 202–229. doi: https://doi.org/10.1525/sod.2021.7.2.202.

Hennings, A., 2019. The dark underbelly of land struggles: The instrumentalization of female activism and emotional resistance in Cambodia. *Critical Asian Studies*, *51*(1), pp. 103–119.

Hiilamo, H., Bitinas, A. and Chân, N., 2020. Extending pension coverage in Cambodia: The governance and investment challenges of the social security investment fund. *International Social Security Review*, *73*(4), pp.97–116.

Ironside, J., 2009. Development-In whose name? In Peter J. Hammer (ed.) *Living on the margins: Minorities and borderlines in Cambodia and Southeast Asia*. Phnom Penh: Center for Khmer Studies, pp. 91–128.

Joshi, S., 2018. *Gendered impacts of land commercialization in Cambodia*. Graduate Institute of International and Development Studies.

Joshi, S., 2020. Working wives: Gender, labour and land commercialization in Ratanakiri, Cambodia. *Globalizations*, *17*(1), pp. 1–15.

Joshi, S., 2020a. *The politics of land grabbing in contemporary Cambodia: Four essays on gender, resistance and state formation*. PhD thesis, Graduate Institute of International and Development Studies, Geneva.

Keeton-Olsen, D., 2021. 'Cambodia puts its arduous process for indigenous land titling up for review'. Mongabay. https://news.mongabay.com/2021/04/cambodia-puts-its-arduous-titling-process-for-indigenous-land-up-for-review/.

Kelkar, G. and Krishnaraj, M. eds., 2020. *Women, land and power in Asia*. Taylor & Francis.

Kent, A. (2016). Conflict continues: Transitioning into a battle for property in Cambodia today. *Journal of Southeast Asian Studies*, *47*(1), pp. 3–23. http://www.jstor.org/stable/43863239. Kieran, C., Sproule, K., Doss, C., Quisumbing, A. and Kim, S.M., 2015. Examining gender inequalities in land rights indicators in Asia. *Agricultural Economics*, *46*(S1), pp.119–138.

Kusakabe, K., 2015, May. Gender analysis of economic land concessions in Cambodia and in Northern Laos: Case of rubber plantations. *LDPI Conference Paper No* (43).

Kusakabe, K., Yunxian, W. and Kelkar, G., 1995. Women and land rights in Cambodia. *Economic and Political Weekly, 30*(43), pp.WS87–WS92.

Lamb, V., Schoenberger, L., Middleton, C. and Un, B., 2017. Gendered eviction, protest and recovery: A feminist political ecology engagement with land grabbing in rural Cambodia. *The Journal of Peasant Studies, 44*(6), pp.1215–1234.

LICADHO, 2014. *'Good Wives': Women land campaigners and the impact of human rights activism*. LICADHO.

Lilja, Mona. 2016. (Re)Figurations and situated bodies: Gendered shades, resistance, and politics in Cambodia. *Signs: Journal of Women in Culture and Society, 41*(3): pp. 677–699. https://doi.org/10.1086/684242.

Maffii, M., 2009. Changes in gender roles and women's status among Indigenous communities in Cambodia's Northeast. In *Living on the margins, Minorities and Borderlines in Cambodia and Southeast Asia*, Centre for Khmer Studies, pp. 129–140.

Murray Li, T., 2014. *Land's End. Capitalist relations on an Indigenous frontier.* Duke University Press.

Neef, A., Touch, S. and Chiengthong, J., 2013. The politics and ethics of land concessions in rural Cambodia. *Journal of Agricultural and Environmental Ethics, 26*(6), pp. 1085–1103.

OHCHR Cambodia Country Office, 2018. *Assessing the impact of social land concession on livelihood in rural Cambodia*. OHCHR.

Park, C.M.Y., 2019. 'Our lands are our lives': Gendered experiences of resistance to land grabbing in rural Cambodia. *Feminist Economics, 25*(4), pp. 21–44.

Park, C.M.Y. and Maffii, M., 2017. 'We are not afraid to die': Gender dynamics of agrarian change in Ratanakiri province, Cambodia. *The Journal of Peasant Studies, 44*(6), pp. 1235–1254.

Peou, C. and Young, S., 2019. Cambodia: Political strife and problematic land tenure. In *Asian Smallholders in Comparative Perspective,* pp. 39–56.

Razavi, S., 2003. Introduction: Agrarian change, gender and land rights. *Journal of Agrarian change, 3*(1–2), pp. 2–32.

Schoenberger, L. and Beban, A., 2020. Rupturing violent land imaginaries: Finding hope through a land titling campaign in Cambodia. *Agriculture and Human Values, 38*, pp. 301–312. https://doi.org/10.1007/s10460-020-10156-z

Thiel, F., 2010. Gender equality and land law in Cambodia. *Facing the Challenges–Building the Capacity*, FIG Congress Paper, pp. 1–15. https://www.fig.net/resources/proceedings/fig_proceedings/fig2010/papers/ts05h/ts05h_thiel_4668.pdf.

Van Der Keur, D., 2014. Legal and gender issues of marriage and divorce in Cambodia. *Cambodia Law and Policy Journal, 2*, pp. 1–22.

Wight, E. 2015. "How land rights are politicising Cambodia's women." *OpenDemocracy.* April 14, 2015. https://www.opendemocracy.net/5050/emily-wight/how-land-rights-are-politicising-cambodia%27s-women

9 Legal pluralism, gender justice, and right to food in agrarian Ghana

Gertrude Dzifa Torvikey and Atudiwe P. Atupare

Introduction

Laws govern norms that in turn determine the distribution of resources, access rules, and legitimizing governance structures. The interconnection between customary and statutory legal frameworks, which is essential in the framework in many rural areas, presents an essential lens through which one can view the realization of the right to food. For women these matter because multiple legal orders can complicate gender justice (Griffiths, 1996). Its complexity is due to the embeddedness of interests and power dynamics deeply rooted in social, political, and economic hierarchies in the society.

As Ubink and Pickering (2020) have noted, an inter-legality framework can block community members from using customary law to redress natural resource disputes while at the same time strengthening the control of the existing institutions. Therefore, the enjoyment of the right to food is locked up in both customary and statutory laws dispensed by multiple institutions. In many countries in Africa, women's access to resources such as land is vexed and constantly challenged and undermined. Meanwhile, in rural areas, the construction of livelihoods is centred on land that is governed by customary laws derived from families, clans, and communities. Community institutions protect men's interest in land while undermining women's land rights. At the same time, as Tsikata and Torvikey (2021) have noted, reforms that seek to re-organize agricultural production do not pay sufficient attention to land and labour dynamics of smallholders. Meanwhile, accounting for this in policy and law is an imperative for ensuring food security and the right to food.

As we have observed in research spanning six years, in agrarian Ghana, disputes arise about women's resource access and use. The land access opportunity of many women is still derived from their association with male relatives as wives, daughters, and nieces. Women are confronting discrimination every day within their families and communities through engaging with plural legal institutions to seek justice, which relates to their access to productive resources and livelihoods with direct bearing on their means of right to food. Drawing from these experiences in the field, we aim to deepen

DOI: 10.4324/9781003202004-13

our understanding of gender justice in connection with the right to food situation within a plural normative framework.

The right to food framework is stated thus:

> The right to adequate food is realised when every man, woman, and child, alone or in community with others, has physical and economic access at all times to adequate food or means for its procurement.
>
> (UNOHCHR, 1999)

It is anchored on five food security frameworks: availability, stability, accessibility, sustainability, and adequacy (FAO, 2014). The comprehensive framework refers to the ability of people to feed themselves from their production and from commons (availability), stability in food situation, the means to procure food without compromising on other needs (accessibility), sustainable management of natural resources to ensure continuous availability of adequate food and the presence of sufficient food required for meeting nutritional needs. The framework has shown that the enforcement of the right to food and food security can be direct and indirect and drawn from international and national normative frameworks. At the national level, individuals or groups must negotiate multiple normative frameworks at the state, community, and family levels to access productive resources essential for sustaining their livelihoods. Multiple legal orders or legal pluralism, which is the coexistence of customary and statutory laws or orders (Griffiths, 1986), is central in gender justice analysis where women have the impetus to choose jurisdiction in matters about their interests. Higgins and Fenrich (2011) argue that legal pluralism presents an opportunity for women to choose jurisdictions. The disproportionate distribution of resources favouring men makes the question of gender justice in rural areas a just one (Seguino, 2013). Gender justice means that men and women should have equal socio-economic rights to live in dignity. Gender justice also envisions equal access to economic resources by women and men and therefore entails the view that women are as agentic as men, and society should concentrate on both equality of opportunities and equality of outcomes. The main questions that animated this chapter include why legal pluralism complicates gender justice and right to food.

The next section contextualizes legal pluralism, gender justice, and the right to food by explaining key concepts. The analytical framework section follows with a description of the typologies of cases and justification of their selection for analysis. This is followed by analysis of the cases, discussion, and conclusion.

Legal pluralism, gender justice, and right to food: a context

Legal pluralism entails the existence of more than one 'legal order', according to which human 'behaviour' is regulated (Griffiths, 1986, p. 1). Legal pluralism can be conceived as the coexistence of statutory and customary law. Ghana is a culturally diverse society with numerous ethnic groups with varied cultural practices and norms, resulting in a plurality of customary codes (Atupare, 2014). Thus, customary rules and practices of the various ethnic communities in Ghana form part of the laws of Ghana (Bimpong-Buta, 1983) and the Constitution recognizes in article 11 the customary laws of the various ethnic groups as part of the corpus of the law in Ghana.

The Constitution provides in article 11 that the 'common law of Ghana' includes the rules of law generally known as the common law and 'rules of customary law', which by custom apply to 'particular communities'. In the light of this, the common law of Ghana embraces not only the general common law but also particular local customary norms (Bimpong-Buta, 1983; Atupare, 2014).

Institutional multiplicity is not only possible by decentralization but is instigated, encouraged, and aided by legal pluralism. Legal pluralism can also ensure local level self-governance through effective negotiation between institutions (Bond, 2008). Multiple normative frameworks combine different positions, interests, power, and authority drawn from local, national, and international levels (Meynen & Doornbos, 2004). Legal pluralism is crucial for gender justice analysis and ascertaining the efficacy of women's right to food and food security in Ghana.

Women in Ghana, whose status is contested, must work and negotiate within these normative frameworks. Therefore, an essential pillar of gender justice encompasses substantive equality principle which is concerned with difference, equitable opportunities, and outcomes.

According to Gheaus (2012), if people suffer injustice because of their sex, they have suffered gender-based injustice. The author further argues that much of the burden of change towards a more just society must be at the individual level. Gender justice emphasizes individual action and agency in achieving gender justice, particularly in the realization of the right to food and food security. However, women who seek justice must confront institutions with varying degrees of understanding and interpretation of women's rights if they recognize it. Women's agency is also crucial and central to the realization of gender-based justices related to the right to food and food security. Although women play vital roles in the realization of households' food security, their relatively low position in society expressed in inequality of enjoyment of political, economic, social, and other rights makes them disproportionately affected by food insecurity (UNOHCHR, 2010).

Women play critical roles in food production as own-account farmers and unpaid workers on household farms. They are also central to food processing and other types of agriculture-based processing that earn income for themselves and their households. In the broader scheme of things, women contribute immensely to producing all sorts of export and industrial crops. As Dzanku, Tsikata, and Adu Ankrah (2021) noted, women account for 43% of the labour on male-held plots and spend 3.9 times more time on reproductive work than men in the household. Yet, due to their discrimination, many women struggle to access productive resources such as land to produce crops that earn higher income in the hierarchy of crops in households.

Apart from this, both productive and reproductive women's labour contribution is less recognized, undervalued, and underpaid (Apusigah, 2009). Similarly, when women produce food crops, most often, their production is on land that is marginalized and already used by male relations, such as fathers, husbands, and brothers. This marginalization of women's roles undermines their livelihoods and negatively affects them, making them poorer relative to men. Meanwhile, the nature of poverty that afflicts women is systemic, and this constrains their capacity to seek justice for such discriminatory practices that affect their access to resources and consequently food security and right to food. These characterizations of the issues that affect the attainment of women's right to food recognize that women are not homogeneous and, therefore, their individual lived experiences are essential in explaining their position in society.

In Ghana, although the legal regime both accommodates and tolerates customary law, aspects of statutory provisions, especially those on family and marriage, have state intervention. Thus, toleration and autonomy interact simultaneously in the agrarian space as communities derive rights and justice from normative frameworks. Shachar (2001) argues that plural legal jurisdictions present women with an opportunity to choose institutions that best protect their rights. Similarly, Bourke Martignoni (2018) adds that there is an opportunity in customary law to achieve the right to food. In addition, the existence of the normative frameworks – customary and statutory – may be reinforcing or could have interactions that will enable women to enjoy their right to food. She further elaborated this point by maintaining relevance for both types of frameworks for the socio-economic realities of rural agrarian areas.

Ekern (2018) holds that different norms and jurisdictional powers regulate social norms. The author explains that legal institutions impact local norms where there is constitutional order and mutuality. But the concentration on conflicts and disputes hides other aspects of the regulating order that shapes individual lives in rural areas, which is expressed in the manifestations of the law and regulations relating to individuals and what individuals do with this manifold of orders. Thus, gender justice pertains to outcomes in access to justice and implications for the right to food operate on different levels and institutions.

Institutional multiplicity can create opportunities for jurisdictional choice and, at the same time, can produce mixed outcomes (Kobusingye, Van Leeuwen & Van Dijk, 2016; Tchatchoua-Djomo, 2018). Institutional multiplicity gives fall-back options to women who must navigate different forms of constraints in their families and communities. The constraints include cost, proximity, and nature of power relations. However, multiple institutions also create and reinforce unjust power relations as various actors compete for recognition and legitimacy.

In Ghana, 78% of the land is held by customary authorities, such as stools, skins, families, and clans (Larbi, 2008), making customary law the main governance framework for land. Similarly, women's labour is central to agricultural production at the household unit. Thus, the realization of the right to food for women and their families is linked to their access to productive and reproductive resources such as land and labour or getting adequate remuneration for their labour to enable them to access food. However, conflicts often arise over land resources, crops, and income from the sale of crops where men control these resources, or over the inheritance of such resources.

Analytical framework

The study draws on qualitative interviews conducted in selected communities in the Eastern, Bono, Northern, and Upper East regions in Ghana. Through our interviews with women, men, community leaders, and representatives of statutory institutions, we found that land, inheritance, spousal property rights, and labour issues dominated. These issues formed the basis of our analysis and have remained salient. They enabled us to reflect on the interplay between legal pluralism, the right to food, and gender justice. We presented the cases thematically and by the various jurisdictional fora used. To present the institutional position on the various themes, we used shared experiences from the community and statutory institutions. In all, we present three vignettes that represent the pattern of cases and their dominance. The Commission on Human Rights and Administrative Justice (CHRAJ) Act, 1993 (Act 456) set up the human rights outfit. Its objective, among other things, is to investigate complaints of violations of fundamental human rights and freedoms, injustice, and corruption, as well as abuse of power and unfair treatment of persons by public officers in the exercise of their duties. It also has the power to seek remedy in respect of such acts or omissions and to provide for other related purposes. Specifically, CHRAJ is required, among other things, to investigate complaints concerning practices and actions by persons, private enterprises, and other institutions where those complaints allege violations of fundamental human rights and freedoms under the Constitution. This open function entitles the Commission to deal with issues bordering on discrimination.

We discussed the church as an emerging forum. We also used a labour dispute between a corporate entity and workers to point to how women's struggles in the communities also tackle corporate entities. We identified various typologies of fora used for the resolution of various disputes, namely (*a*) family unit, (*b*) a chief's court, (*c*) church, (*d*) statutory court, and (*e*) CHRAJ.

We used narrative accounts where the women themselves narrate their struggles. In addition, we used direct quotes for institutional accounts about cases that concerned women. To show the typologies of women's encounters within a pluralistic legal framework, we followed the routes of the cases as the women narrated them. This approach helps us understand what determines the jurisdictions women choose, particularly when they start with one at the initial stages.

Women's legal struggles, right to food, and outcomes

Women in the study communities used three prominent institutions, namely the traditional courts (family and traditional courts), statutory courts, and CHRAJ, to adjudicate disputes. Dispute resolution through the church emerged as the fourth forum used. As the cases will demonstrate, the women generally start with traditional dispute resolution institutions such as the family and resort to statutory institutions when the case is unresolved.

The family unit is mainly the first port of call, and cases are moved to the traditional apex courts when the issues remain unresolved or there are unsatisfactory outcomes. Even though customary institutions are inherently biased against women, their proximity, power, and initial cheaper dispute adjudication cost make them the preferred choice for the women we interviewed in the communities. In one of the study communities, a woman spent GHC 50 (USD8.37) for a hearing in a traditional court. Her opponent, who lost the case, refunded the GHC50 the woman spent on the legal processes. The opponent also paid the traditional court GHC50 for the entire process and provided local drinks. This cost of justice delivery in the chief's court is relatively cheaper compared to legal fees charged by lawyers. According to the Ghana Legal Council (GLC), approved legal fees for junior counsel is GHC500 (maximum) and hourly rates for Ghanaian clients is GHC300–1,000 (USD50–166). And for land litigation, the cost is '10%–20% of the value of the land or hourly rates would apply if the work involved exceeds 20% of the value of the land'. Similarly, the charges for human rights cases ranged from 'GHC6,000-30,000 [USD993.55–4967.73] or hourly rates would apply if the hourly rates involved exceeded GHC30,000'.[1]

Traditional courts are also a pragmatic choice because women can relate to customary laws and regulations and articulate them with little second-party interpretation. The traditional courts also have a compelling force since they draw legitimacy from the people and are well recognized. Disputes on women's disinheritance and dispossession are prominent in the traditional dispute resolution fora. The following description is an example of an outcome of a case a woman sent to the chief's court:

> A man died, and his family disputed the widow's claim to the cocoa farm she and her late husband had established. The family tried to dispossess the woman, and she sent the case to the chief's court. The court resolved the dispute and allocated part of the farm to the widow.
>
> (Assemblyman, Kasapin, Asunafo North District, 22 February 2016)

In this case, where the opponent is her late husband's family, the woman chose the chief's court as a neutral arbiter. By winning the case and getting part of the cocoa farm, the traditional court has invoked tradition that advances the woman's right to the cocoa farm, which is central to her livelihood and advancing her right to food. Apart from the income that she will earn through the sale of the cocoa beans, the woman can also cultivate food crops for sale and household consumption. Additionally, she can use the farm as collateral for credit, which will also enhance her livelihood. However, it is unclear the level of fairness that the court advanced to the women in this case.

Sometimes, women's choice of multiple customary institutions depends on the patriarchal structure they confront, which also presents challenges that demand women negotiate with different institutions and change strategies. The following case is an example of cases going through multiple institutions to varying degrees of success and failures.

Land and cocoa farm inheritance dispute (woman vs her brother)

> Maame Afua was a 30-year-old farmer. She completed Junior High School and is married with a child. Her parents migrated from Aburi to Asumura in the Bono region a very long time ago. They acquired vast lands for cocoa farming. Maame Afua was born in Asumura.

She narrates land and cocoa farm dispute between her and her brother that is over a decade old:

> My father was chief before he died, and my brother inherited the throne. Before my father died, he bequeathed me land and a cocoa farm. My brother disputed my inheritance of the sharecropped cocoa farm our father left for me. I sent the case to our family head which I won. But my brother did not obey the ruling. I reported the case to community leaders, but my brother was the chief. I would not say I liked the way the traditional court dealt with the case. So, I involved our church in impressing him to respect the earlier ruling from the family, and he stopped disturbing me. During the seven years of the dispute, I was given the proceeds of the cocoa farm. So, I reclaimed the cocoa farm. As soon as he agreed to let go of the cocoa farm, he shifted his attention to the ten acres of land my father left for me. The land our father left for him is in the same area. I insisted that he could not take my land from me because that was my land. I used that piece of land to cultivate food crops to feed myself and my family. I also sold some of the crops in the market. However, my brother continued to harass my farmworkers and me. One day, he sent thugs to chase them away from the farm. He claimed that my father left the land for him and not me. During the period of the dispute, my brother gave parts of the land to sand winners and took the proceeds. I again sent the case to the chief's court, but my brother refused to leave the land. I tried planting pineapples on the land, and he sent land guards to uproot the suckers. Then I tried planting cocoa on the land, and he

> again stopped me. He also threatened me, and because of that, I stopped pursuing the case so that he would not kill me. I feared for my life.
>
> (interviewed on 20 February 2016, Asumura)

Maame Afua used community institutions, namely the family, the chief's court, and the church, to varying degrees of success. Maame Afua's brother drew from his authority as a chief, also the higher-order institution in the community, to frustrate the case. The multiple jurisdictions present an opportunity for Maame Afua to choose and change strategy. With her brother being the chief, it became clear why Maame did not proceed with the case in the traditional court because she would not get justice from that avenue. She decided to use the church, which combined religious and cultural normative frameworks to redress the cocoa farm dispute but not that of the land. With the exhaustion of the multiple customary adjudicating jurisdictions, she stopped pursuing the case and surrendered the land to her brother. Maame's actions could be described as the choice of jurisdiction; however, her agency was done with socio-economic calculations that have consequences for her. It is important to note that, while Maame and her farmworkers were constantly harassed by her brother (the chief), she did not report the assault case to the police in the community. While some scholars have argued that women could choose jurisdiction, the freedom to choose is also constrained by a huge social cost and the nature of patriarchy that the woman must confront. In Maame's case, her brother is also the chief and, should she push the case out of the community to a statutory institution (such as the courts, the police, and CHRAJ), it might mean she was up in arms against the chieftaincy institution and the community itself rather than her brother. If this happened, she would have to contend with the social forces in the community. Choosing the state over the community might mean that she was subjecting the whole community to the dictates of statutory laws. Meanwhile, leaving the land case could afford her the opportunity to belong to the community and benefit from its resources.

As has been shown in Maame Afua's case, the question remains as to what happens when the opponent disrespects the ruling or chooses another dispute resolution institution that disfavours the woman socially and economically or even politically? The following cases demonstrate this critical question,

> I had a land boundary dispute with a young man. I shared the boundary with his mother. For the past three years, I could not get to the end of the boundary since I could only weed a small piece at a time. It will take me five years to weed the entire land. Three years ago, I was informed that the son of the woman I shared the boundary with had taken part of my land and was cultivating it. I confronted him, but he did not stop, and so I informed his mother. The young man did not listen to his too. I then reported him to his family, but he refused to take part in the case resolution. I was advised to send him to the chief, and when the chief called

> him, he ignored him. To my surprise, the young man reported the case to the Koforidua High Court, and till now, the case has not been heard.
> (Adwoa, 97 years, widow, Middle School Leaver, mother of 7 children, Abaaem, Kwaebibirem District, 25 February 2016)

In Adwoa's case, her opponent was unwilling to submit himself to the jurisdiction chosen by her and therefore dragged her to court, the forum he chose. Adwoa exhausted all possible customary avenues with little success. Her opponent ignored traditional institutions and chose the statutory courts instead. Adwoa has already indicated resource constraints as having prevented her from cultivating the entire land at once. Most courts are in districts and regional capitals and not physically accessible to rural dwellers. As this case has shown, the jurisdiction an opponent chooses and finds legitimate could affect a woman's resilience in fighting for justice. The case has also delayed and Adwoa might find it harder to continue the case.

Both Maame and Adwoa's cases have illustrated some aspects of chieftaincy and the status of chiefs that have come under scrutiny. While some argue that the role of chiefs is not anachronistic to a modern state (Abotchie, 2006), others highlight their weakening power particularly in land matters as they have been accused of inherent biases (Ubink, 2007; Campion & Acheampong, 2014).

Some women also rely on their knowledge of customary and statutory laws to pursue justice, even when they are resource-constrained. Maame Afriyie's resilience in a disinheritance case is illustrative as she stated:

> I know that if someone bequeaths property to another party before they die, then no one can lay claim to that property. Recently a male family member summoned me to court claiming the lands that I inherited from my uncle. I spent GHC 3,000 [USD 504] to resolve the case. It became a big case, and I eventually won. I took a loan to settle the debt we incurred at the court. I got my land back but had a debt to repay.
> (Widow, 90 years, Bomso, Kwaebibirem District, 19 February 2017)

There were also labour disputes in the study communities. For example, the following describes a labour dispute involving female workers in a processing firm whose termination of appointment led to a court case.

Labour dispute (female workers vs company)

> A group of women employed on contract to work in a processing facility in the Kwaebibirem District summoned the company to court for terminating their contracts. In the first instance, the women obtained an injunction in court against the company to stop producing. However, the company got this injunction overturned on appeal. It continued to produce. The case is still in court for the past three years.
> (Company official, Abaaem, 24 February 2016)

There appears to be a distinction between corporate entities and those involved in non-corporate entities. The workers chose to send a formal entity to court for infringement on their right to work and violation of the Labour Act, 2003 (Act 651). Since the remit of their employment is governed by statutory law (Act 651), the court is best suited to resolve the labour dispute. However, delays in the judicial process constrained the outcome of the case. The case also shows that women in the community are not only fighting patriarchy but also capitalism.

The Commission on Human Rights and Administrative Justice

In the study communities, CHRAJ reported having dealt with two prominent cases about women. These were disputes on the dispossession of women, particularly widows and divorced women's joint property ownership. In agrarian Ghana, Nyantakyi and Kerr (2017) found pervasive dispossession of women, especially of land, and this takes two forms: gradual and systematic. For married women who relied on land allocated to them by their husbands, the authors argue that when there is resource scarcity, the presence of a husband guarantees the security of the land. It means that, for instance, even if the man has not formally allocated a piece of land to his wife, the land tenure for the woman is insecure. However, if the man lives in the house with his wife, then the latter's land use is a bit secure. However, when the man relocates or dies, it renders the ownership and control of the resource insecure.

Widow dispossession, a human rights violation, is a common customary practice in many parts of rural Africa. Although the nature of marriage determines the widow's security of the property, it is argued that the widow's inheritance of a deceased husband's property is still contingent on the widow's good disposition and the ability of her husband's family to see her good behaviour (Willis, 2018).

In Ghana, the Intestate Succession Act, 1985 (PNDCL 111) was enacted to correct the dispossession of living spouses and children when the deceased dies intestate, which means that in the legal order, statutory law clashes with family law in inheritance issues. CHRAJ reported many cases of dispossession of widows when families of a deceased husband prevent the widow from taking charge of the resources she and her husband had worked to acquire. The human rights institution uses its discretion based on the type of case and the parties involved to protect the women's interests in the property. Officials of CHRAJ explain how they resolve disputes of this nature. An official stated:

> In most cases that come to our office, women are denied access to land at the demise of their husbands. Usually, the head of the extended family, clan heads or brothers-in-law are the opponents in such cases. Therefore, settlement depends on who is involved. If it is the brothers-in-law, the family head is usually invited to settle the case through mediation.
>
> (CHRAJ, Garu Tempane District, 19 February 2017)

> A widow reported her late husband's family to us for disinheriting her. We invited both parties. We listened to the merits of the case. Accordingly, we applied PNDCL 111 to secure her interest in the property. She got her share of the property eventually. CHRAJ receives many cases of this nature from the District. Many people come here to complain about losing their relatives and other relatives preventing them from inheriting what is rightfully theirs. It is so especially with women who lose their husbands. They have their husband's family preventing them and their children from inheriting the deceased man's assets. We usually invite both parties and settle the cases.
>
> (CHRAJ, Kade, 22 February 2016)

> A woman came to report her ex-husband about the land they jointly acquired in Kumasi. The man was developing the land without the woman's consent. They are indigenes of Bawku, but the land acquired was not in the Garu Tempane District; we directed the woman to send the case to CHRAJ in Bawku. We received information that CHRAJ ruled that the land should be divided into two so that the woman can have her share.
>
> (Social Welfare Officer, Bawku, 18 February 2016)

The three cases above indicate the varied ways CHRAJ dispenses justice to women. In the case of dispossession of women at the demise of their husbands, the human rights institution uses PNDCL 111 to settle the cases, although the administration of justice in this manner rests in the bosom of the courts. CHRAJ can only achieve this through reliance on some of the traditional institutions, particularly the family. Although CHRAJ draws its powers from its statutes, adjudication of the inheritance issues mainly relies on informal powers if the families agree to settle the case with CHRAJ. Therefore, we can see CHRAJ as a bridge between traditional and statutory institutions, which shows the constant negotiation between many legal orders (Sieder & Barrera, 2017). Hammond (2019) argued that the concessional approach to adjudication of property-sharing cases is more practical and can reduce conflict and legitimacy disputes. The approaches that CHRAJ uses also raise questions about the hybridity rather than binary legal orders since customary institutions are also consulted in these cases. As the cases have shown, the human rights institution has become a famous statutory dispute resolution centre for many women as many realized it takes less time than the courts. On average, a case may take a maximum of three weeks to be resolved. CHRAJ does not charge fees for their services. Since the officials at the office are not necessarily lawyers, the institution refers the more complicated cases to the courts. While CHRAJ applies PNDCL 111 widely in cases of widow dispossession, that law has been critiqued for weaknesses that do not account for equality as enshrined in international human rights laws that enjoin states to eschew discrimination against women (Gedzi, 2014). Currently, an amendment of the law is pending before the country's parliament.

Discussion and lessons

All cases analysed illustrate gross violation of women's rights, including the right to inherit and to control jointly acquired property, which fundamentally undermines their socio-economic rights and right to food. However, we have also shown women's struggles to assert their rights. Maame Afua has reclaimed the inherited cocoa farm but lost the ten-acre land that she was using to produce food, cocoa, and other cash crops. Both her survival and accumulative strategies will be hampered by the loss of land. Adwoa's land has been blatantly encroached upon by a young man who is willing to take her through a long legal battle. Similarly, the labour case has been delayed and the women continue to suffer the unjust lay-off.

Women's social locations within rural areas dominated by customary law and the distance to statutory institutions determined their experiences with normative pluralism. The cases that we have discussed show that legal pluralism, on the surface, shows that women can have the freedom to choose. However, beneath this pluralism are power relations that mediate women's choice of jurisdiction in the countryside such that women may choose the jurisdiction that has less financial and social cost to seek justice (Whitehead & Tsikata, 2003). Other considerations may include the effectiveness of the chosen jurisdiction. The presence of multiple jurisdictional institutions in rural areas reflects the legal structure of the country. It also provides a window to understand how women navigate the different fora to seek justice to advance their socio-cultural and economic rights and, by extension, right to food.

We have also highlighted class, knowledge of the laws and institutions, access to resources including land and finance, and social proximity to institutions as necessary for women's justice-seeking behaviour. These categories of difference have analytical power as we have shown that these sometimes condition whether a woman seeks justice, continues to seek justice and the outcomes of the cases sent to specific jurisdictional fora chosen. Similarly, opponents also determine the legal path by choosing a jurisdiction that might be unfamiliar to women. Since disputes involve people and parties, we also paid particular attention to the forms of patriarchy that women confront and contest on an everyday basis.

Aligning with Griffiths (1986), who argues that the state in Africa has limited control over land matters in rural areas, it is not surprising that many women sent cases to traditional institutions. It is essential to recognize that women know the limitations of the statutory courts and laws, and therefore they negotiate with customary institutions. However, the refusal of some opponents to respect the ruling or submit themselves to the traditional courts shows the weaknesses emerging in the traditional jurisdiction systems. Manji (1999) argues that state laws do not secure women's interest in legal pluralistic frameworks and that one must look at other ways women's mistrust of the state are manifested and how they outsmart the legal systems. As our cases have shown, CHRAJ, a statutory institution, has been used by some women to 'outsmart' customary institutions and statutory courts and minimize costs and the incumbrances of these institutions. Legal pluralism enables one to see both community and statutory governance frameworks, allow women to negotiate with patriarchy, and advance their rights within customary and statutory normative frameworks. In rural areas where land is crucial for the construction of livelihoods, while its governing institution remains male and women suffer in its distribution, legal pluralism becomes essential for women to negotiate with institutions that provide immediate remedies for them. Women's institutional choices, even within constraints, amplifies the multiple strategies they use in dealing with the state and customary institutions.

Conclusion

Women's access to productive resources and recognition of their labour in the household is mediated by socio-cultural norms, further transformed by the history of commodity production. With the changes in gender relations over time, the family has become a centre for struggles for gender equality in access to and distribution of resources. Consequently, the Constitution grants some recognition to customary laws while, at the same time, common laws prevail.

Statutory laws regulate aspects of customary practices that discriminate against women. Hitherto, customary laws were used, especially in cases where women's inheritance and land rights are concerned. The passage of PNDCL 111 has provided a new legal hierarchy for adjudicating on inheritance and property rights. In this chapter, we have problematized the position of choice of jurisdiction, which has characterized the discussion about women's experiences of multiple normative frameworks. Although women have multiple jurisdictions that they can use, we have argued that they are constrained by political, economic, and social hindrances. We found that the type of patriarchy women confronts, the jurisdiction chosen by their opponent, the nature of the case, and the cost and time of delivery determine how they seek justice and persist in pushing for rights that affect their right to food. As the cases have shown, while some women may mobilize resources to pursue cases, and incur debt in the process, their resistance to patriarchal oppression shows their quest for justice to enhance the protection of their livelihoods. Women's relative social position in society is a hindrance to seeking justice in the law courts. Going to statutory courts requires money, time, and energy, especially since many courts are mainly located in urban areas where many lawyers practice. A judge in the Garu Tempane district reiterated this point about the social encumbrances that confront women as they seek justice:

> Even in court here, a woman is thought of as chattel. She is part of the property the man owns. That is why women are unable to take advantage of the law due to how society positions them.
>
> (Judge, Garu Tempane District, 19 February 2016)

This statement reflects the customary realities of women. We should note that dispute resolution in statutory and traditional courts have come under heavy criticism for several reasons, and these have implications for women and their quest to seek justice. For example, widespread public perception of judicial corruption has further deepened the negative perceptions about the judicial processes and the courts. While this leaves the traditional courts with some leverage in settling disputes, they too have been critiqued for gender bias in their composition, procedures, and processes (Gedzi, 2009). Consequently, CHRAJ and the church have become intermediary dispute resolution institutions at the local level. However, these must also take a concessionary approach to dispute resolution by weighing the case, drawing from custom and statutory normative frameworks. While CHRAJ, a statutory human rights organization with legitimacy from the state, draws from international and national human rights domains and transmits these to the local level, their approaches to resolving the issues remain informal and can be contested. Therefore, the success they report depend mainly on the parties' cooperation, which shows that some compromises are made. CHRAJ had recognized this weakness when it indicated that complex inheritance disputes are referred to the courts. Since normative pluralism is conflictual and leaves room for dilemmas, legitimacy is a key element for institutions within the normative framework to function properly (Wolf, 1994; Macdonald & Macdonald 2019). CHRAJ has used its force as a statutory organization with human rights principles to apply in Ghana's context gender equality and justice in tandem with the statutory laws as the cases have shown. However, only women who send cases to CHRAJ benefit from its jurisdictional approach. The right to food framework shows that one needs resources to produce or acquire food. For women in rural areas, where land is a crucial resource, struggles with patriarchy and customary institutions over their land rights enable them to assert and reclaim their right to food. Similarly, women are also confronting corporate entities that deny them rights to decent working conditions to ensure that they have money to buy food or invest in economic activities that ensure their food security and right to food. While access to productive resources is essential for realizing human rights such as those on the right to food, courts in Ghana emphasize civil and political rights over socio-economic rights (Atupare, 2011, 2014), although this is crucial in ensuring the right to food.

Note

1 https://www.glc.gov.gh/resources/gba-approved-fees/.

References

Abotchie, C. (2006). Has the position of the chief become anachronistic in contemporary Ghanaian politics? In Odotei, I.K & Awedoba, A.K. (eds.). *Chieftaincy in Ghana: Culture, governance and development*. Accra: Sub-Sahara Publishers, pp. 169–191.

Apusigah, A. (2009). The gendered politics of farm household production and the shaping of women's livelihoods in Northern Ghana. *Feminist Africa*, 12(12): 51–67.

Atupare A. P. (2011). Legitimacy, judicial review, and human rights enforcement in Ghana. *University of Ghana Law Journal*, XXIII, 228–310.

Atupare, A. P. (2014). Reconciling socio-economic rights and directive principles with a fundamental law of reason in Ghana and Nigeria. *Harvard Human Rights Journal*, 27(1), 71–106.

Bimpong-Buta, S. Y. (1983). Sources of law in Ghana. *Review of Ghana Law*, 15, 129–151

Bond, J. E. (2008). Pluralism in Ghana: The perils and promise of parallel law. *Oregon Review of International Law*, 391(10), 391–418.

Bourke Martignoni, J. (2018). Engendering the right to food? International human rights law, food security and the rural woman. *Transnational Legal Theory*, 9(3–4), 400–414.

Campion, B. B. & Acheampong, E. (2014). The chieftaincy institution in Ghana: Causers and arbitrators of conflicts in industrial Jatropha investments. *Sustainability*, 6(9), 6332–6350.

Dzanku, F. M., Tsikata, D., & Adu Ankrah, D. (2021). The gender and geography of agricultural commercialisation: What implications for the food security of Ghana's smallholder farmers? *The Journal of Peasant Studies*, DOI: 10.1080/03066150.2021.1945584

Ekern, S. (2018). Between relations and rights: Writing constitutions in Mayan Guatemala. *The Journal of Legal Pluralism and Unofficial Law*, 50 (2), 167–187.

FAO. (2014). The right to food within the international framework of human rights and country constitutions. Rome: FAO.

Gedzi, V. S. (2009). Principles and practices of dispute resolution in Ghana: Ewe and Akan procedures on females inheritance and property rights. Dissertation Submitted to the International Institute of Social Studies of Erasmus University Rotterdam, The Hague, The Netherlands.

Gedzi, V. S. (2014). PNDC Law 111 in Ghana and international human rights laws. *Global Journal of Politics and Law Research*, 2(2), 15–26.

Gheaus, A. (2012). Gender justice. *Journal of Ethics & Social Philosophy*, 6(2), 1–24.

Griffiths, J. (1986). What is legal pluralism? *Journal of Legal Pluralism and Unofficial Law*, 18(24), 1–55.

Griffiths, A. (1996). Legal pluralism in Africa: The role of gender and women's access to law. *Polar*, 19(2), 93–108.

Hammond, A. (2019). Reforming the law of intestate succession in a legally plural Ghana. *The Journal of Legal Pluralism and Unofficial Law*, 51(1), 114–139.

Higgins, T. & Fenrich, J. (2011). Legal pluralism, gender, and access to land in Ghana. *Fordham Environmental Law Review*, 23(2), 7–21.

Kobusingye, D. N., Van Leeuwen, M., & Van Dijk, H. (2016). Where do I report my land dispute? The impact of institutional proliferation on land governance in post-conflict Northern Uganda. *The Journal of Legal Pluralism and Unofficial Law*, 48(2), 238–255.

Larbi, W. O. (2008). Compulsory land acquisition and compensation in Ghana: Searching for alternative policies and strategies FIG/FAO/CNG International seminar on state and public sector land management Verona, Italy, September 9–10, 2008.

Macdonald, T. & Macdonald, K. (2019). Towards a 'pluralist' world order: Creative agency and legitimacy in global institutions. *European Journal of International Relations*, 00(0), 1–27.

Manji, A. S. (1999). Imagining women's 'legal world': Towards a feminist theory of legal pluralism in Africa. *Social and Legal Studies*, 8(4), 435–455, 010357.

Meynen, W. & Doornbos, M. (2004). Decentralising natural resource management: A recipe for sustainability and equity? *European Journal of Development Research*, 16(1), 235–254.

Nyantakyi-Frimpong, H. & Kerr, R. B. (2017). Land grabbing, social differentiation, intensified migration, and food security in northern Ghana. *The Journal of Peasant Studies*, 44(2), 421–444.

Seguino, S. (2013). Toward gender justice: Confronting stratification and unequal power. *Géneros*, 2(1), 1–36.

Shachar, A. (2001). *Multicultural jurisdictions: Cultural differences and women's rights*. Cambridge: Cambridge University Press.

Sieder, R. & Barrera, A. (2017). Women and legal pluralism: Lessons from indigenous governance systems in the Andes. *Journal of Latin America Studies*, 49, 633–658.

Tchatchoua-Djomo, R. (2018). Improving local land governance? Exploring the linkages between land governance reforms, institutional pluralism, and tenure security in Burundi. *The Journal of Legal Pluralism and Unofficial Law*, 50(1), 31–55.

Tsikata, D. & Torvikey, D. G. (2021). Rural women's livelihoods and food security in Africa. In Berik, G. & Kongar, E. (eds.), *The Routledge handbook of feminist economics*. London and New York: Routledge, pp. 216–224.

Ubink, J. (2007). Traditional authority revisited: Popular perceptions of chiefs and chieftaincy in peri-urban Kumasi, Ghana. *Journal of Legal Pluralism*, 39(55), 123–161.

Ubink, J. & Pickering, J. (2020) Shaping legal and institutional pluralism: Land rights, access to justice and citizenship in South Africa. *South African Journal on Human Rights*, 36(2–3), 178–199.

United Nations Office of the High Commissioner for Human Rights. (2010). The right to adequate food. Factsheet No. 34: Geneva: UNOHCHR.

Whitehead, A. & Tsikata, D. (2003). Policy discourses on women's land rights in Sub-Saharan Africa: The implications of the re-turn to the customary. *Journal of Agrarian Change*, 3(1/2), 67–112.

Willis, R. (2018). A comparative analysis of widow dispossession in francophone and Anglophone Cameroon. *Journal of African Law*, 62(1), 147–174.

Wolf, S. (1994). Two levels of pluralism. *Ethics*, 102(4), 785–798.

10 Social security in the extractive state

Gender, land inheritance, and agrarian change in Ratanakiri, Cambodia

Alice Beban and Joanna Bourke Martignoni

Introduction

Land inheritance has traditionally been a key social protection mechanism in the rural world. Young people are able to start or consolidate their own farms through inheritance, while older people secure care by providing land to children who look after them. Even though the shift from subsistence to commercial farming means that farming is becoming less important than wage labour to household incomes, land still retains a crucial affective, social, and economic function within rural families. In countries where rapid agricultural commercialization and 'land grabs' have radically altered the landscape, however, smallholder farmers face land scarcity. Since the mid-1990s, the Cambodian government has actively pursued agricultural development objectives underpinned by an 'extractive' logic within which private economic interests and patronage networks reinforce centralized authoritarian political power. Wealth and natural resources have been concentrated in the hands of the elite without any mechanisms to promote egalitarian resource redistribution. Thus, at the same time as it is undermining informal, family-based social protection, the state continues to be conspicuously absent as a provider of social security.

As land becomes a scarce commodity, land-poor families in Cambodia's Northeast are increasingly concerned with negotiating inheritance for their children to guarantee economic and social security for all members of the family unit. While the growing literature on land relations in Cambodia tends to focus on conflict between households and agribusiness/political elites, every day, intimate struggles over resources also need to be recognized (Brickell, 2020; Jackson, 2003). Families negotiate inheritance decisions in diverse ways, taking into account intersecting demands and expectations over old age care, familial love, inter-generational wealth, fairness, and affective relationships with the land that show a complex and dynamic moral economy of land transmission. In this chapter, we examine changing inheritance practices as a lens through which to understand the gendered and

DOI: 10.4324/9781003202004-14

generational effects of the incursion of capitalist social relations in rural areas.

This chapter is based on extensive qualitative interviews carried out by members of the DEMETER research team in 2016–2017 and again in 2019–2020 in Ratanakiri province, Cambodia. Ratanakiri is home to the majority of the country's indigenous groups and increasing numbers of Khmer (ethnic majority) who have migrated from lowland provinces. We draw from around 180 semi-structured interviews and ten focus group discussions with farmers (women and men), including indigenous Charay and Tampuan ethnic groups, as well as Khmer. All interviews were conducted in Khmer or Tampuan and Charay languages, and transcribed into English. The transcripts of the semi-structured interviews have been coded in NVivo using inductive coding. In what follows, we first discuss theories and relevant legal provisions on social protection and the moral economies of land inheritance; then we examine changes in inheritance practices at the household level, and discuss how these practices contribute to shifting social relations of class, gender, age, and indigeneity to produce new forms of empowerment and inequality in rural communities.

Social protection and the absent–present state

Social protection[1] consists of practices, policies, and programmes aimed at reducing vulnerability and enhancing capacity to manage risk in times of social and economic crisis (ICRW, 2006). Although the right to social security is clearly articulated in Article 22 of the Universal Declaration of Human Rights (1948) and in Article 9 of the International Covenant on Economic, Social and Cultural Rights (1966), in many countries, social protection is increasingly being privatized within families as states following neo-liberal approaches have retreated from active provisioning. Nancy Fraser (2016) describes how the current, financialized form of capitalism is 'systematically consuming our capacities to sustain social bonds' by withdrawing public social supports and depending on families (particularly women) to provide unpaid care as well as income from paid work. This has resulted in a 'crisis of care', which is intertwined with the ecological crisis, as capitalism sees both nature and social reproductive capacities as free gifts to exploit. Even where formal social protection schemes are provided, rural people who work on their own farms, do caring work, and labour in informal employment are often ineligible for the state benefits that accrue to urban people in formal employment (Cook et al., 2003). Social security systems are also premised upon gendered constructs with a presumptive male breadwinner and, as in the case of the Cambodian constitution, women being positioned as 'housewives' (ICRW, 2006; Sepúlveda Carmona & Donald, 2014). The burden of caring work usually falls disproportionately on women, whose invisible reproductive labour, alongside structural discrimination in education and employment, excludes them from high-earning economic opportunities and formal employment relations (Goldblatt & Rai, 2020; ICRW, 2006; Sepúlveda Carmona & Donald, 2014).

The ability to access and control land is a strategy used by rural people to manage economic shocks and social risks including family dissolution, illness, disability, and indebtedness. Land rights for men and women are often affirmed in legislation, although customary law regarding land inheritance usually prevails on the ground and may, in some contexts, perpetuate inequalities of gender, age, class, and ethnicity (Toulmin, 2009). Across Asia, the growing dominance of globalized markets has reshaped the social contract between state and citizens in relation to access to land (Rao, 2008). This has gendered implications: market mechanisms often favour private capitalists over family-based small farms, and even if legal and policy frameworks support equitable land rights, neo-liberal economic policies use state power to legitimize the alienation of communities, especially women, from their rights to land and resources (Rao, 2008).

In Cambodia, Article 36 of the constitution states, 'Every Khmer citizen shall have the right to obtain social security and other social benefits as determined by law'. The 2001 Land Law makes provision for gender equality through presumptive joint land registration and titling (Thiel, 2010). The Civil Code, adopted in 2007, provides that both women and men have equal rights in accessing bank loans and inheritance.[2] The 2019 Law on Social Security aims to entrench a system of comprehensive maternity, unemployment, illness, disability, and old age pension coverage that would be extended to those working in the private sector as well as civil servants who were covered under the pre-existing benefits schemes defined in the 2008 National Social Security Fund for Civil Servants (Hiilamo et al., 2020). In practice, however, these state schemes remain inaccessible for the majority of rural people, particularly women who tend to have less of the education and political connections that would enable them to access formal employment in the public sector. This situation compounds pre-existing gender inequalities (ILO, 2012). The state may therefore be present as a provider of social security in policy documents and legislation, but this has little meaning for rural people if they are not aware of these policies, or if they deem the judicial and administrative systems untrustworthy, inaccessible, or violent and therefore do not claim their entitlements. Amongst people we spoke with, there was rarely any sense that the state should be providing health, education, adequate housing, or social security in the event of disability, illness, old age, or caring responsibilities.

The state thus exists as a 'ghostly' absent presence in the everyday lives of rural upland communities (Gordon, 2008); absent in terms of public social service provision, but increasingly present as a mediator of global capital and an enforcer of private property rights for those with the right political connections and economic power. Upland spaces have rapidly been brought into global circuits of capital and into the everyday bureaucratic realm of the state. Starting from the mid-2000s, and particularly around 2012–2013, agribusiness concessions, land speculators, and lowland migrants have moved into the province *en masse*, encouraged by state economic development policies promoting agribusiness investment through Economic Land Concessions (ELCs), and the provision of private land titles. Communities that previously saw land as a plentiful and fertile common resource have rapidly developed a sense that it has become scarce and impoverished. The idea that land is essential to life was a dominant theme in our interviews, as one woman eloquently stated:

> If we have land, we can work on it for survival. But if we do not have land, we sell labour and just earn income from day to day. How can we survive? People hope on land to improve their lives and their children's, on and on… As a mother, I worry about land.
>
> (44-year-old woman, November 2019)

The arrival of ELCs had a profound impact on people's farming and inheritance practices. Communities had to clear and cultivate whatever land they could to prevent the land being grabbed by companies and well-connected elites. Families that lacked the labour and resources necessary to quickly clear land, or who sold land due to debt or hardship, were forced to turn to low-wage labouring. Park and Maffii (2017, p. 1248) suggest that a 'gender fissure' is emerging in indigenous communities as men have more mobility than women. Young men are more likely to speak Khmer and to have access to mobile phones and motorbikes, while young women face the cumulative obstacles of time scarcity, workload and responsibilities as farmers and carers, and the barriers created by patriarchal norms and discrimination. However, literature on agrarian transformation makes clear that we cannot assume any simplifications about the ways commercialization of agriculture affects people, and upland women have diverse aspirations, class positions, and integration into commercial agriculture and off-farm livelihoods (Frewer, 2017).

Changing moral economies of land inheritance

The burgeoning literature on land in Cambodia tends to focus on land acquisition through the state and market or on land conflict between village households and agribusinesses (Sovachana and Chambers, 2018; Un and So, 2011). Until recently, there has been little discussion of intra-familial land inheritance practices and the way in which macroeconomic structures are influencing the everyday struggles and land inheritance strategies within extended families[3] (FAO, 2019). What we find in practice is a diversity of mechanisms through which families negotiate various intersecting demands and expectations over old age care, affective relationships, and intergenerational wealth and fairness that show an economy of land inheritance based on complex moral judgements. The notion of a 'moral economy' of land inheritance draws attention to the mutuality of relationships involved in these transactions that are not exclusively economic in nature but also repose upon social and cultural values (Carrier, 2018; Gudeman, 2016). The outcomes of these negotiations are often filled with ambiguity for land-poor women.

State intervention through the formalization of land tenure rights may be beneficial for women where legislation and policies mandate equal inheritance and rights over land in cases of divorce or widowhood. In practice, however, in many countries, the formalization of land rights has instituted a patriarchal system of titling land in the names of male heads of households, which can deprive women of their traditional property rights (Peters, 2010). In the case of Ratanakiri, our research has documented many cases of departure from the formally equal land inheritance rights laid out in legislation in the interests of preserving matrilineal land inheritance customs that seemingly favour women. Amongst groups with communal land governance and use structures, such as pastoralists and indigenous communities, privatization of communal land can undermine customary land tenure arrangements. Lesorogol points out that this does not necessarily dissolve existing relationships, however; rather, it can create a gap in social institutions relating to land inheritance that encourages a diversity of practices to proliferate (2014). This is certainly true in Ratanakiri where the complex mix of collective forest, fishery, and grazing lands that was previously managed by indigenous communities together with their family farms and residential land has increasingly been enclosed and privatized. This has occurred despite the constitutional recognition of indigenous collective title and the mechanisms supposedly in place to enable indigenous communities to assert claims over communal land (Bourdier, 2019).

The growth of diverse non-farm livelihoods in Southeast Asia's rural communities means that there is no longer a straightforward linkage between land inheritance and gendered social status. Land inheritance may restrict people's mobility and ability to gain education and off-farm livelihoods or to 'utilise land through markets' (Fernando, 2015). In communities such as those in upland Cambodia where women may have previously secured status through land inheritance, increasing land atomization, whereby inherited plots are too small to be economically viable, may also be eroding the function previously performed by land in the provision of social security (Agarwal, 1991).

In rural areas, land access and ownership has traditionally played a key role in a person's social status and economic security. A rich literature in feminist economics and agrarian studies confirms how important land rights are for rural people's food security, women's autonomy, and sense of identity (Quisumbing, 2003; Tsikata & Golah, 2010). Access to and control over land is fundamentally gendered, and patterns of inheritance favour men in most of the world (Cheryl Doss et al., 2014; Deere & Leon, 2003). However, differences among women are equally important (Doss et al., 2014). Women's changing relationships – whether as members of landholding families, as members of groups that do not own land, as migrants, and as divorcées or widows – structure their access to and control of land. Within Southeast Asia, the predominantly bilateral and matrilineal inheritance systems in land are said to give women more status than in South Asia, with its patriarchal systems of land control (Kieran et al., 2015). It is important to remember, though, as Errington (1990) points out, that while bilateral and matrilineal inheritance may provide status, this does not mean women have *control* over the land (Peluso and Lund, 2011) or that they are able to exercise each of the different 'bundles of rights' attached to ownership (Doss & Meinzen-Dick, 2020), nor does it mean that women have access to property management and public authority (Agarwal, 1994).

Inheritance norms are not static; as agricultural production practices become more market-oriented, there is a tendency for matrilineal societies to adopt the more restrictive practices of patrilineal inheritance (Lastarria-Cornhiel, 2005), undermining women's customary land rights (Jacobs, 2009b). But feminist scholars are careful to note that customary arrangements do not always give women more rights of access and control, and claims to land must be constantly negotiated (Jacobs, 2009a). State efforts to formalize land rights also influence customary systems in ambiguous ways. Formalizing land rights can be beneficial to women when states mandate equal social protection provisions in patriarchal systems, such as bilateral inheritance and women's control over land in cases of divorce/widowhood (Jacobs, 2009b). But, in practice, the formalization of land rights often institutes patriarchal systems of titling land in male head of households' names, and in matrilineal societies, mandating bilateral inheritance harms women's status (Peters, 2010).

The growth of non-agricultural livelihoods in rural communities calls into question the relationship between land inheritance and gendered social status. As societies adapt to land scarcity and market conditions, land can paradoxically become both more valuable (as scarcity causes land prices to rise), and also assume less importance in rural household livelihood strategies as off-farm work becomes more prevalent (Rigg, 2006). In contexts where the contribution of agriculture to household incomes has declined, women's land rights may increase their work burden without providing them with greater status or decision-making authority in their households and communities (Rao, 2006). Land inheritance can restrict people's mobility and ability to gain education and off-farm livelihoods, 'shackling' young people to the soil (Fernando, 2015). While the ability to inherit land can secure some status for women, this might no longer be the case once inherited plots become too small to be commercially productive. Agarwal (1994), however, argues that even if land is not large enough to be economically viable it still plays a role as part of livelihood diversity, but this requires empirical understanding of how this works as part of diverse household strategies, and careful consideration of how land may increase the double/triple burden on women as well as increasing their status.

'The lucky and unlucky daughter': moral economies of land inheritance in Ratanakiri

The legal provisions governing inheritance practices in Cambodia – in particular the Civil Code – assert that parents should divide assets equally amongst their children. But our study revealed a huge diversity in inheritance norms, and no one felt that the law reflected their reality; in fact, one young man noted, 'In our country we follow the parents first and then the law, but for other countries they comply with the law first and then parents' (22-year-old Charay man, April 2016). All indigenous Charay families said that the traditional practice in their communities was to transmit land and the family name to their daughters (Park & Maffi, 2017).

Customary land rights comprise a mosaic of residential and farm lands that passed through the family line, and community forest land that young couples could clear for practising shifting cultivation. Most people felt it was appropriate to give land to daughters and not sons, because matrilocal norms meant that sons would gain land through marriage and men were seen to have the strength to clear forest land if needed: 'Daughters do not know how to get land. For men, after their marriages, they will come to live with the lady's side and get land' (35-year-old Charay man, November 2019). Some families passed land on equally amongst daughters upon the daughters' marriage, while families with less land often held back plots for the daughter who would care for them in old age (usually the youngest daughter). This custom means, as one 17-year-old indigenous woman said, that the youngest daughter is the 'lucky and unlucky daughter':

> She is lucky because she will get the property more than any other daughter in the family even the home not only land. She is unlucky because when parents get older and cannot do farming anymore, she will take all responsibilities to taking care of them and feed them.
>
> (17-year-old Charay woman, March 2016)

Indigenous Tampuan families in our study had different inheritance norms from Charay families; while still following a matrilineal inheritance line, most families gave some land to sons also. Amongst the Khmer, inheritance practices were heterogeneous. Most practised bilateral inheritance, with all children receiving farmland, and holding back residential land for the person who would look after their parents in old age – usually the youngest daughter (60-year-old Tampuan woman, April 2016).

Traditional Charay and Tampuan matrilineal inheritance practices provide a higher status for women than in patrilineal communities. With no dowry to pay and with the woman bringing land into the relationship as well as carrying on the family name, many people said that daughters were favoured because they continued the family line (40-year-old Charay woman, April 2016).

But now that land is increasingly scarce, being the provider of land is a double-edged sword for women's status: women are rooted to the land, while men are more mobile. In fact, many families said they did not pass on land to their sons, but they passed on assets such as 'money, motorbikes, and gold necklaces', which provide men with the mobility to access education and job opportunities outside the village, and money to make different livelihood choices (52-year-old Charay woman, March 2016).

Changes induced by land scarcity

In Ratanakiri, the moral economy of land inheritance is changing, as people no longer have enough land to pass on to children and there is very limited access to forest for clearing new land. We found that, rather than any strong new norms emerging, social norms are in flux, as people find different strategies to cope with the scarcity, trying to help their children get established, and prepare for old age. We analyse these emerging norms below through a generational lens, looking first at the implications for youth and then for their parents.

Transmission of land to the younger generation

Matrilineal or gender-neutral inheritance?

Land scarcity and the loss of forest land means that families with sons face the possibility that their daughter-in-law's family will not have land to pass down to her, and their son will become landless. Families had different ways of dealing with this possibility. Some people planned to continue their matrilineal inheritance traditions and said their son would just have to make his own way (65-year-old Charay man, April 2016). Others suggested that the old inheritance norms were no longer fair if men were unable to inherit land from in-laws or clear new land plots from the forest. For this reason, families were starting to pass small land parcels down to their sons and daughters. Many people shared a sentiment that sons also should now get land, particularly among families that possessed enough land to be able to spare some for sons. A typical case is one Charay woman whose parents distributed land only to their daughters in the customary way, but who is now married with two sons and one daughter, and is considering passing some land on to all her children when they marry:

> I will consider it when they marry. If their wives inherit land, I will give [my sons] small lands such as 1ha each. The rest, I will inherit to my daughter, and my husband and I will live with my daughter until the end of my life. If their wives only inherit a small amount of land, I will consider giving my sons more land.
>
> (38-year-old Charay woman, April 2016)

The prospect of marrying into a family that did not have land worried both parents of sons and the young men themselves. In one interview with a

55-year-old Charay man and his son, both spoke about their anxiety for the next generation:

> SON: If parents do not pass on land to sons, how can they have land for living? If my parents do not inherit land to me, everything will be finished. If a son gets married to a lady whose parents also do not have land, what could he do? From where could he find land in this time?
>
> FATHER: I agree we should inherit land to all children. It is not like the past where people could clear land wherever they wanted. I lie down and I think of the next generation; they face difficulty from today onwards.
>
> (24 year-old and 49 year-old Charay men, March 2016)

Like others in formerly matrilineal families, this father had passed farmland on to all his children, and had also divided his large residential land plot to allow his children to build houses next to his, a shift in norms that disrupts long-held matrilocal norms in this community whereby married couples live near the wife's natal family. This change has implications for childcare and support, as the wife's mother often takes a large role in assisting her daughter as she raises her own family, and this informal social security may not be available to her if daughters relocate to live with their husbands upon marriage.

Land use rights and shared cultivation

This context of shifting inheritance norms has thrown customary practices and social status linked to land into flux. Families who provide land in their son's village may have to act cautiously to avoid shaming his wife's family, or they may worry that they will not have enough land to divide when their younger children marry. People are devising creative strategies to deal with the new context of land scarcity, including providing sons and their spouses with use rights to the land rather than passing down ownership (62-year-old Charay woman, November 2019). Some indigenous families said they share cultivation of rice with their parents and parents-in-law because the rice land is not large enough to divide between children, and then the wider family shares the rice harvest among themselves for eating; this arrangement is more unusual with cash crop fields. This practice of giving use rights rather than ownership caused tension in some families; one Charay woman in her 50s with eight children said she gave use rights to her oldest children to share management of her land rather than passing some on, because she did not have enough for everyone: 'I have so many children … [and] my land is not enough to distribute to my children… Now, my daughter complains about her father who does not inherit her the land yet' (55-year-old Charay woman, March 2016).

Children choosing not to take inherited land from parents

Adult children also exercised agency to refuse land from their parents. Some said they chose to give back their land or not take land their parents promised them because they knew that this would leave their parents in a difficult position in future years (22-year-old Tampuan man, February 2020). Others said it was inconvenient to take the land as they did not live nearby, or they would buy land themselves (32-year-old Tampuan woman, March 2016). This care for the family extended to the wider family unit. For example, a Khmer man who was brought up by his grandmother inherited land and livestock from her and from his parents. When his in-laws offered him and his wife land, they refused because they felt that her parents did not have enough land to survive themselves. 'This is because my side we have enough land and some property but my wife's family have nothing … they gave us the land but we rejected it because I know they don't have anything' (23-year-old Khmer man, November 2019).

Delaying inheritance and marriage

Some families who found themselves suddenly facing land scarcity – whether because of land grabs, forest enclosure, or land sales – did not yet have a strategy for passing on their assets, and many seemed unsure what they would do. In some cases, families were putting off passing on land, even though their children had already married. Some young people said they were delaying getting married due to a lack of money and land to farm and build a house on. One Charay woman in a relationship said she was currently growing cassava to save money, and she hoped to get married soon because, 'If we do not do it, I am afraid that the ghost or spirit will come' (22-year-old Charay woman, February 2020). Some families said they were using family planning methods to have fewer children, because, as one woman noted, 'I don't want to have more children because … there is no land for the next generation to access' (30-year-old Charay woman, April 2016).

Changing notions of valuable inheritance land

With the need for cash to pay for food (that now must be bought from the market due to the reduction in rice production and forest loss, as well as the growing desire for, and status associated with packaged food), and money to pay for farming inputs, school fees, and other new 'needs', the notion of what value inherited land has is also changing. Passing on uncleared forest land was seen as less desirable than passing on land with mature fruiting cashew trees. Even if parents had very little land to pass on to children, growing cashew on the land could increase its value for children through providing a steady income source. In one interview, a woman commune councillor instructed another Charay woman to grow cashew on the land so her children could gain ongoing income:

> Growing cashew nuts is good for children. When they marry, they can make money from selling cashew nuts. You have only one daughter and two sons. This would not be difficult for you.
>
> (38-year-old Charay woman, April 2016)

Value also depended upon what crop 'boomed' at the time; one Khmer woman said that while she only received one hectare of land from her parents, she received a mature rubber field while her siblings received cassava fields, and were forced to pull out the cassava after the price plummeted (39-year-old Khmer woman, February 2020). Cashew nut trees were also valuable, and could provide an income for children and grandchildren, as children could harvest the nuts. One young Charay woman from a large family of eight siblings said her father was determined to provide an income for his grandchildren so they did not have to work for others:

> My father grows cashew trees for me [on] land which he inherited to me. He said if he passes away, at least there are cashew nuts for his grandchildren to harvest and sell to get cash … All of these lands, there are people asking to buy but my father said he won't sell land. … Land now is very expensive. Like my land, if I decided to sell, they would give me $30,000, but my father said he doesn't want to sell because he has many children and grandchildren.
>
> (18-year-old Charay woman, April 2016)

This quotation is interesting also for the debate it raises over what can be done with inherited land. While we heard from many people who sold some or all of their inherited land when land prices rose, this woman seemed tempted to sell but suggested it was her father's decision.

Care for the older generation

Cambodian parents making land inheritance decisions are concerned both with how to give their children the best start in adult life, and how to ensure that they themselves are cared for in older age, in a context where no formal systems of social support are available. Older people who do not have land to share with their children said they were fearful that they would not be cared for, such as a Charay woman who said:

> I used to feel so happy because I thought I had big land for my children but now I don't know what to think because if I share to my children, I will have no land when I am in old age. When I am old and I still have land, my children will come close to me and take care of me but if I don't have land when I am old, no one will come to see me.
>
> (58-year-old Charay woman, March 2016)

There was a strong sense from this woman and others that caring for parents was a moral obligation expected of children, but also that it was not automatic. People talked about the possibility that their children may choose not to care for them, and what they would do in this situation:

> If children love parents, then they would give some money to parents. But if they do not love parents, they won't give money to parents.
>
> (40-year-old Charay woman, December 2019)

Widowed or divorced women without children, or whose children are estranged, are particularly vulnerable if they have limited land resources and often have to rely on extended family networks for support. In one case, an indigenous woman married a Khmer migrant who sold land she had inherited from her own family, then left her and went back to Phnom Penh. His brother, who continued to live locally, said that when she was older, his children would look after her. In other cases, husbands had retained land upon divorce, sometimes selling it, which meant that the children could not receive land. The father of one divorced woman with three children said that,

> All three children are living with my daughter, and her husband did not inherit any property to the children. The village chief said that I can file a complaint but now it is over.
>
> (55-year-old Charay man, November 2019)

In cases where people did not have daughters who would care for them, most Charay people said the custom was for the niece (child of wife's sister) to inherit land and come to take care of people in their old age. But some disagreed, and one man said this was a 'stupid idea' because only their own children would take care of them, so they would rather pass land down to their son and his wife even if this meant doing things differently. When the interviewer expressed surprise that this man did things differently from other interviewees, the man said he was influenced by Khmer families:

> I started to think like this about three or four years ago. No reason we inherit our land to the other beside our children. Just following the culture without thinking properly I think it's wrong. I met and discussed with Khmer people, they also said that 'whether son or daughter, they are [still] our blood'. We have to inherit to them equally. I will not inherit my land to other people that will not take care of me and my wife when we are old. Currently, I still have power to work so it is ok but when I am getting old, how can I survive? If I get sick, who can help me? There are only my children.
>
> (55-year-old Charay man, April 2016)

This quote suggests that the emerging norm of inheriting land bilaterally to both daughters and sons appears to be motivated both by the possibility that sons will not receive land from their in-laws, as well as the influence of Khmer migrant practices of bilateral inheritance. One 25-year-old Charay woman expressed this as: 'Charay people learn from Khmer practice. They are afraid that sons will be jealous of daughters who inherit land. So some families start passing on land to sons too' (25-year-old Charay woman, March 2016).

A second way in which Khmer migrants influence inheritance norms in indigenous communities is through the growing number of mixed marriages between Khmer and indigenous people. Most often, this was Khmer men marrying indigenous women. Some indigenous interviewees felt that Khmer migrant men wooed and married indigenous women just so they could claim land from them. Some indigenous families responded by changing their practices when their daughters married Khmer, such as several indigenous women who said their mothers did not share any land with them upon marriage as they did not trust their Khmer husbands:

> I do not have land. I only work on my mother's small land ... She did not dare inherit land to me with a land title because I got married to a Khmer man. She is afraid that my husband would sell that land. So I just keep working on her land and get some harvest for sale.
>
> (32-year-old Charay woman, April 2016)

Strategies for ensuring income and care in old age

Several families who did not have land to bequeath to their children said they hoped to buy land in order to pass it on, on the understanding that this would mean that they would be cared for once they were no longer able to work. Others tried to secure more land by selling the small plots they inherited and buying land in more remote areas that was cheaper so that they could give their children a farming livelihood and continue to live with them. This was a risky strategy, however, as more remote land was difficult to access and maintain. Some older people whose children did not live nearby said they hired others to work on their land if they could afford it. One older Charay man who grew cashew, banana, and pepper, said 'I earn money from renting my ten hectares of farming land to buy rice to eat... Both of us are old enough to relax' (58-year-old Charay man, April 2016). The differences in these strategies depended fundamentally on class divisions, as the new pressures on land had deepened inequalities of land ownership within and between ethnic groups.

Outcomes of new pressures on land: deepening class, gender, and ethnic divisions

One of the main implications of land scarcity and the changing patterns of land inheritance is a deepening of class, ethnic, and generational divisions within communities. People's narratives identified growing inequality between those who were able to amass land (and capital through logging) prior to land prices rising and forest land enclosure, and those who had little or no land to pass on to their children and few other decent livelihood options available.

This is particularly an issue for girls, as their status has previously been reproduced as holders of the family name and land, but they get married early and usually receive limited schooling. This means it is difficult for them to move into jobs that require literacy, and they are more likely to end up labouring on other people's farms if they do not have sufficient land in their own family. When we asked indigenous families who had limited land how their children would cope, they often connected land scarcity to the growing class of landless labourers in the village, and the potential that their children would become labourers:

> I do not know what we will do [about land inheritance]. We will wait and see. Now, parents could inherit around one hectare to each of their children. And it is hard to live with this amount of land. They would live by selling labour their whole life.
>
> (42-year-old Charay man, March 2016)

Khmer families who had not managed to amass land holdings were similarly concerned about their ability to provide for their children, and worried that their children would become labourers or migrants to the city or to Thailand. Others suggested that children who did not have land to inherit might have to delay the transition to adulthood and forming their own household, because if they did not have land 'they can only continue living with their parents and taking care of their parent's farm, tapping their parents' rubber or going to work for other people' (30-year-old Khmer woman, March 2016).

Labouring for others was seen as a shameful activity, of a lower status and less secure than owning one's land. However, this sentiment was expressed strongly in our first round of interviews in 2015–2016, while in 2019–2020, fewer people discussed the shame of labouring. While almost all interviewees still desired to own land, people noted that in the past year, labourers could do better than farmers because of a fluctuation in crop prices and weather, and because there is now such a divide between those who own a lot of land and others who own just small plots and work for others.

Out-migration

The strategy often followed in situations of land scarcity in lowland Cambodia – migration out of the community, either temporarily or permanently – was evident amongst Khmer families. As land has become scarce and the logging trade has been progressively closed down, some Khmer migrant families who have not managed to secure long-term land are leaving the area, sometimes selling land and at other times retaining fallow lands for their children to inherit. However, out-migration has not yet become a dominant livelihood strategy amongst most indigenous families. Women, in particular, are less mobile. This is the other side of having status as the holder of the land in the female line: women are seen as less able to drive a motorbike, to travel far into the forest or into the city. So the men are the ones who have taken up job opportunities in mines or plantations; men are much more likely to have an education, which can potentially set them up for decent off-farm livelihoods, and a few men are now migrating to Thailand or Phnom Penh, as is common amongst lowland Khmer.

Those labouring opportunities that are available locally are often low paid, and the rate of remuneration is generally lower for women than men. Because most labour is paid piecemeal (depending on performance), older people and those who are unable to work as fast receive less pay, and some told us that they are simply not considered for work. It is uncommon in some communities for indigenous people to labour on the large plantations. They are more likely to work for other indigenous families, and sometimes for Khmer migrants, but the availability of these jobs fluctuates.

Some mothers talked about the need to begin focusing on educating their daughters as well, since they could not rely on having land, and education was seen as a potential path out of poverty and of the dependence on labouring for others. Most families, however, did not talk about strategies for preparing their daughters for off-farm employment, and while a lot of families hoped to give their daughters more education, this was limited by the inaccessibility of schools (high schools were far from the community, which meant children had to stay in the district town, which was expensive and some families saw this as inappropriate for girls) and gender norms that reproduce the notion that girls should marry young. And many young people were well aware that they may not receive land from their parents, but in the absence of any decent off-farm livelihood options they knew of in the area, and with out-migration not (yet) a feature of life for the Charay communities, they had a sense that they did not know what to do. The interviewer asked a 15-year-old girl who was sitting listening to the adults talking about land scarcity:

Q: So, in your generation, how could you have land for production?
A: Maybe there is no land inheritance from my parents.
Q: So, are you currently learning any professional skill or are you at school?
A: I do not learn any skill and I also dropped out of school. (15-year-old Charay girl, March 2016)

Conclusions: land, care, and social security in the absent–present state

In Ratanakiri, the complex mosaic of communal and individual land rights that, in the past, has been reproduced and supported through mechanisms of matrilineal land inheritance has been disrupted through the workings of neo-liberal agricultural development and the adoption of private and joint land titling schemes (FAO, 2019). Our research shows that access to communal forest and grazing land, as well as the social security that comes through matrilineal land inheritance patterns, are particularly important for women and girls, as well as for elderly family members, and that land continues to perform a crucial affective and cultural function in upland communities, in addition to being an economic asset.

The role played by land as a form of informal social security is, however, filled with ambiguity, and possession of land can serve to limit the mobility of women or even lead to abuse and harm by those in positions of decision-making power in relation to land inheritance. In addition, the specific vulnerability of older people who are no longer strong enough to engage in farm labour and therefore not viewed as productive members of rural communities is clear in our interviews.

The situation our interviewees describe in this chapter reveals the 'absent presence' of the state in rural Cambodia, where the state actively reshapes rural communities through economic development policies favouring large agribusiness and commercial farming practices, while the provision of care and protection is left largely in the hands of families and communities. This 'absent presence' is by no means unique to Cambodia. A rich literature shows how people's interactions with the state depend on subjectivities formed through gender, rurality, ethnicity, age, and other contextual social factors (Williams, 2011). This notion of the state as 'absent presence' also works on an analytical level to go beyond the common trap of work on plural/hybrid systems and neo-patrimonial politics that sees the state in the global south as always what it is *not* (a rational, legal bureaucracy with independent judiciary) to the exclusion of what it *is* – 'the myriad ways in which politicians, state makers, and different sectors of society make claims and engage with each other' (Strauss & Cruise O'Brian, 2007, p. 2). Recognizing that the state is neither simply present nor absent directs the researcher to work through the ways that the state is both markedly marginal in the regulation of daily life and areas in which it is directly active, and to attend to moments in which shifting relationships between state and society re-shape informal social protection mechanisms. In this chapter, we have shown how state and corporate actions in the 'public' domain of economic development and land governance influence the 'private' sphere of intra-familial land inheritance practices and informal social protection across generations.

The Cambodian case highlights broader spatial and temporal phenomena linked to globalized, neo-liberal capitalism that promotes commoditization of agriculture and land while simultaneously constraining social protection. Rural people actively strategize to ensure care for themselves and their children in this context of land scarcity, and we see new moral economies of land inheritance emerging based on bilateral rather than matrilineal inheritance, and on a revaluing of off-farm livelihoods and assets. These strategies may enable individual families to shore up care for themselves and their children; however, they also reinforce growing class and gender inequalities in rural communities. In the absence of legal and policy frameworks to ensure state-provided social security, these transformations are likely to leave women from poorer families with reduced social status and limited mobility, while condemning many of the next generation of smallholder farmers in Ratanakiri to a life of landlessness.

Notes

1 Social protection can be 'formal' protection provided by the state and market (private insurance firms), or various 'informal' individual and collective care arrangements, and sometimes a combination of each of these.
2 This is enshrined in Article 1156: 'Children are first in line for succession. They receive equal shares regardless of whether they are natural or adopted', and Article 1162, 'The spouse of the decedent is always entitled to succession. If there are children, the surviving spouse receives an equal share'.
3 A 2019 paper on land inheritance and migration, for example, bases its description of inheritance practices and family formation on May Ebihara's well-known anthropological study from 1968 and a small study from 2001 (Takahashi, 2001). This tendency to justify contemporary assertions about inheritance practices using a limited selection of historical works tends to reproduce stereotypes about the conventional patterns followed, with a common-sense assumption that Khmer families follow bilateral inheritance and indigenous families follow matrilineal inheritance patterns. See the discussion in FAO (2019, p. 13).

References

Agarwal, B. (1994). *A Field of One's Own: Gender and Land Rights in South Asia*. Cambridge University Press.
Bourdier, F. (2019) From Confrontation to Mediation: Cambodian Farmers Expelled by a Vietnamese Company. *Journal of Current Southeast Asian Affairs*, *38*(1), pp. 55–76. https://doi: 10.1177/1868103419845537.
Brickell, K. (2020). *Home SOS: Gender, Violence, and Survival in Crisis Ordinary Cambodia*. Wiley.
Carrier, J. G. (2018). 'Moral Economy: What's in a Name'. *Anthropological Theory*, *18*(1), 18–35. https://doi.org/10.1177/1463499617735259.
Cook,S.,Kabeer,N.,&Suwannarat,G.(2003).*SocialProtectioninAsia*.HarAnandPublications. https://www.researchgate.net/publication/47651591_Social_Protection_in_Asia
Deere, C. D., & Leon, M. (2003). The Gender Asset Gap: Land in Latin America. *World Development*, *31*(6), 925–947. https://doi.org/10.1016/S0305-750X(03)00046-9
Doss, C., & Meinzen-Dick, R. (2020). Land Tenure Security for Women: A Conceptual Framework. *Land Use Policy*, *99*, 105–180.Doss, C., Summerfield, G., & Tsikata, D. (2014). Land, Gender, and Food Security. *Feminist Economics*, *20*(1), 1–23. https://doi.org/10.1080/13545701.2014.895021
Errington, S. (1990). *Power and Difference: Gender in Island Southeast Asia*. Stanford University Press.
FAO. (2019). *Women's Land Rights and Agrarian Change: Evidence from Indigenous Communities in Cambodia*. www.fao.org/
Fernando, A. N. (2015). Shackled to the Soil: The Long-Term Effects of Inherited Land on Labor Mobility and Consumption. *SSRN Electronic Journal*. https://doi.org/10.2139/ssrn.2693327
Fraser, N. (2016). Contradictions of Capital and Care. *New Left Review*, *100*, 99–117. https://newleftreview.org/issues/II100/articles/nancy-fraser-contradictions-of-capital-and-care
Frewer, T. (2017). The Gender Agenda: NGOs and Capitalist Relations in Highland Cambodia. *Critical Asian Studies*, *49*(2), 163–186, https://doi.org/ 10.1080/14672715.2017.1300861

Goldblatt, B., & Rai, S. (2020). Remedying Depletion through Social Reproduction: A Critical Engagement with the United Nations' Business and Human Rights Framework. *European Journal of Politics and Gender, 3*(2), 185–202.

Gordon, A. (2008). *Ghostly Matters: Haunting and the Sociological Imagination*. University of Minnesota Press.

Gudeman, S. (2016). *Anthropology and Economy*. Cambridge University Press.

Hiilamo, H., Bitinas, A., & Chan, N. (2020). Extending Pension Coverage in Cambodia: The Governance and Investment Challenges of the Social Security Investment Fund. *International Social Security Review, 73*(4), 97–116.

ICRW. (2006). *Property Ownership & Inheritance Rights of Women for Social Protection - The South Asia Experience*. https://www.icrw.org/wp-content/uploads/2016/10/Property-Ownership-and-Inheritance-Rights-of-Women-for-Social-Protection-The-South-Asia-Experience.pdf

ILO. (2012). *Social Protection and Gender in Cambodia*. Evaluation Summary. ILO Evaluation Unit. https://www.ilo.org/wcmsp5/groups/public/

Jackson, C. (2003). Gender Analysis of Land: Beyond Land Rights for Women? *Journal of Agrarian Change, 3*(4), 453–480. https://doi.org/10.1111/1471-0366.00062

Jacobs, S. (2009a). *Gender and Agrarian Reform*. Routledge.

Jacobs, S. (2009b). Gender and Land Reforms: Comparative Perspectives. *Geography Compass, 3*(5), 1675. https://doi.org/10.1111/j.1749-8198.2009.00257.x

Kieran, C., Sproule, K., Doss, C., Quisumbing, A., & Kim, S. M. (2015). *Examining Gender Inequalities in Land Rights Indicators in Asia*.

Lastarria-Cornhiel, S. (2005). *Gender and Property Rights within Post-conflict Situations*. www.dec.org

Park, C., & Maffi, M. (2017). 'We Are Not Afraid to Die': Gender Dynamics of Agrarian Change in Ratanakiri Province, Cambodia. *The Journal of Peasant Studies, 44*(6), 1235–1254. https://doi.org/10.1080/03066150.2017.1384725

Peluso, N. L., & Lund, C. (2011). New Frontiers of Land Control: Introduction. *Journal of Peasant Studies, 38*(4), 667.

Peters, P. E. (2010). Our Daughters Inherit Our Land, But Our Sons Use Their Wives' Fields: Matrilineal-Matrilocal Land Tenure and the New Land Policy in Malawi. *Journal of Eastern African Studies, 4*(1), 179–199. https://doi.org/10.1080/17531050903556717

Quisumbing, A. (2003). *Household Decisions, Gender and Development*. IFPRI.

Rao, N. (2006). Land Rights, Gender Equality and Household Food Security: Exploring the Conceptual Links in the Case of India. *Food Policy, 31*(2), 180–193. https://ideas.repec.org/a/eee/jfpoli/v31y2006i2p180-193.html

Rao, N (2008). *'Good Women Do Not Inherit Land': Politics of LAND and Gender in India*. Routledge. https://www.routledge.com/Good-Women-do-not-Inherit-Land-Politics-of-Land-and-Gender-in-India/Rao/p/book/9781138501928

Rigg, J. (2006). Evolving Rural–Urban Relations and Livelihoods in Southeast Asia. In C. Tacoli (Ed.), *The Earthscan Reader in Rural–Urban Linkages* (p. 68). Earthscan.

Sepúlveda Carmona, M., & Donald, K. (2014). What Does Care Have To Do with Human Rights? Analysing the Impact on Women's Rights and Gender Equality. *Gender and Development, 22*(3), 441–457. https://doi.org/10.1080/13552074.2014.963305

Sovachana, P., & Chambers, P.W. (2018). Human Insecurity Scourge: The Land Grabbing Crisis in Cambodia. In C. Hernandez, et al. (Eds.), *Human Security and Cross-Border Cooperation in East Asia* (pp. 181–203). Palgrave Macmillan.

Strauss, J., & Cruise O'Brian, D. B. (2007). *Staging Politics: Power and Performance in Asia and Africa*. I.B. Tauris & Co.

Thiel, F. (2010). Gender Equality and Land Law in Cambodia. *Facing the Challenges–Building the Capacity*, 1–15.

Toulmin, C. (2009). Securing Land and Property Rights in sub-Saharan Africa: The Role of Local Institutions. *Land Use Policy*, *26*(1), 10–19. https://doi.org/10.1016/j.landusepol.2008.07.006

Tsikata, D., & Golah, P. (2010). *Land Tenure, Gender, and Globalisation: Research and Analysis from Africa, Asia, and Latin America*. ZUBAAN; International Development Research Center.

Un, K., & So, S. (2011). Land Rights in Cambodia: How Neopatrimonial Politics Restricts Land Policy Reform. *Pacific Affairs*, *84*(2), 289–308. https://doi.org/10.5509/2011842289

Williams, P. (2011). An Absent Presence: Experiences of the "Welfare State" in an Indian Muslim Mohalla. *Contemporary South Asia*, *19*(3), 263–280.

11 Constitution, courts, right to food, and gender equality in Ghana

Atudiwe P. Atupare

Introduction

In 1992, the political leadership of Ghana made a decision either unilaterally or popularly consulted and were either forced or persuaded to return the country to a constitutional democracy. The primary rules of politics at the time were largely the enactments of the military government. But constitutional democracy, as the reports of history and contemporary practice will have us believe, cannot be run on such military laws. A new constitution, it was rightly thought, must be made through a deliberate act with the required legitimacy to constitute the fundamental starting point. It would thus seem that the act of constitution-making was inevitable for Ghana. This constitution was supposed to mirror the collective aspirations and values of the people. The true reflection of these values in the new constitution would make the people the authors of such a [legal document] as constitution.

So, though the act of constitution-making is a pre-eminently political act (Elazar, 1985), it entails, in our view, more than that as it sets out and defines rights and rights holders, and legal duties and duty bearers in a state.

> It is a decision-making process carried out by political actors, responsible for selecting, enforcing, implementing, and evaluating societal choices; and it is shaped by the socio-political order in which it takes place, and, in turn, it strongly influences that order.
>
> (Kornblith, 1991, pp. 61–89)

The standing argument has been that the stability of a constitution remains a characteristic aspiration: drafters intend to set values and institutions for generations to come. This does not only remind the framers of an intergenerational value concern, but reminds the whole nation that a stable constitution might be one that takes into account legal values such as rights.

In this chapter, we assess the protection of the right to food and gender equality in Ghana. Our basic claim is that whereas the Constitution of Ghana is less explicit about the protection of the right to food, the right to gender equality is expressly set out. Be this as it may, the courts of Ghana over the

DOI: 10.4324/9781003202004-15

years have shown a positive and receptive attitude towards this mixed textual position by protecting both the values of the right to food and gender equality. Nonetheless, such a judicial reception has not been fully embraced and benefitted from by the people in the area of study. Findings of the research show acts that are contrary to the basic elements of the right to food and gender equality.

Constitution-making: weak constitution with strong court

In making a constitution, the framers may work towards or uphold a purpose of providing for a document that will serve as the principal driving force for collective aspirations of the people. That is, the orientation of the constitution being drafted or made is determined by its actual or contemplated role and the values, either parochial or international by nature, which the constitution must or is envisaged to protect. At the general level, we may have among others the purpose of nation-building as a new state emerges: the consolidation of democracy as the military retires to the barracks or authoritarian presidents are deposed; liberalism and the creation of private markets with the end of communism; peace and cooperation among communities to end internal conflicts. Each of these generalities must be fed with the peculiarities or specific context, history, aspirations, or hopes of the state.

In the case of Ghana, the history of military dictatorship, egregious violation of human rights, poverty, neglect, misery, and social decadence conduce the constitution-making purpose to partly protecting fundamental human rights and freedoms. There is an expectation of a constitution that will mirror both the national and international appetite for the respect, protection, and fulfilment of the core values that human rights entail. The collective hopes and aspirations of the nation will be seen to be promoted, respected, protected, and fulfilled with a constitution that takes the case for human rights seriously. Thus, the legitimacy of the constitution is tied to the liberal framework and substance of the human rights provisions of the constitution. Such a constitution will not be seen as primarily and almost exclusively a legal document only for rules of law and for practically nothing else, but also it will be seen as a sort of manifesto, a confession of faith, a statement of ideals, a 'charter of the land' (Wheare, 1966).

In that context, a constitution may be measured for its worth or strength based on its human rights provisions. In post-conflict or authoritarian situations, a constitution is thought to be strong if it has adequate provisions reflecting all the rights and freedoms that the global human rights constituency argues for. Where there are few or insufficient provisions on rights, the constitution is adjudged as a weak one. So, a weak constitution, it is argued, blunts the hopes of human rights advocates, courts, and the people of that state. The situation will deepen their misery and neglect and provide no distinction in terms of constitutional values and political experience between their previous lives and what is provided for in the post-conflict or post-authoritarian state. This is even complicated with a near cancerous hopelessness where there is a weak or positivist judiciary that might exclusively focus on a mere logical disquisition of rights from the poor provisions of such rights that exist in the constitution.

However, this view of desolation may not be entirely true; a weak constitution on rights may not necessarily obfuscate all opportunities on human rights enforcement. In fact, a strong court can do well for the human rights constituency in that state with a weak constitution. Put differently, a strong court with the appropriate judicial methodology can do well with a weak constitution on human rights. Thus, the poverty of human rights or their values in a national constitution is not necessarily an unrevised judgment of despair for human rights claimants in that country. Such a fate only obtains where there is a weak constitution with a weak court. In any case, where both the constitution and the courts are weak, the view is that the peoples' constitution-making objective is defeated and the struggle for good constitutional governance must continue.

The general constitutional environment of Ghana

Ghana is a republic governed under a constitutional democracy. At present, the sources of law in Ghana are, among others, the 1992 Constitution (hereinafter referred to as the 'Constitution'), Acts of Parliament, Common Law of Ghana, Orders, Rules, Regulations and Customary Law as applicable in the various local communities (article 11). The Constitution in a self-proclaimed tone stated that, in the hierarchy of legal norms, it is the supreme law of the country (article 1(2)). We may say that the Constitution provides the rule of recognition or the general test or criteria of legality for all laws applicable in the state.

The Supreme Court affirmed this vision in 1996 when it held it as 'the bulwark which not only fortifies the supremacy of the Constitution but also makes it impossible for any law or provision inconsistent with the Constitution to be given effect to'. Such a position adopted by the court suggests that the doctrine of parliamentary supremacy, as is sometimes referred to in the United Kingdom, does not exist in Ghana's legal system. Parliament as an authority or institution may exercise its legislative powers either in making new laws or amending or repealing existing laws subject to the strictures imposed by the Constitution.

A point worth stating here is that article 11 does not expressly mention international law as part of the sources of the laws of Ghana. That is, if we are to stay with this provision for the sources of law or applicable legal rules in Ghana, there is nothing explicit on international law as a source of law. However, it has been observed that a liberal construction of this provision contemplates within its meaning that international law is a source of law in Ghana. This is even more feasible if there is a structural reading of articles

11 and 75 together. As a matter of principle, article 75 vests in the President of the Republic the power to execute or to cause to be executed treaties, agreements, or conventions in the name of Ghana, subject to ratification by an Act or resolution of Parliament. As it would seem, international law in the form of treaties, conventions, and protocols may become part of the laws or applicable legal rules of Ghana if the appropriate constitutional procedure is adhered to.

The international law and municipal law dichotomy

Since international law has a place in the state of Ghana as legal rules, it is essential to settle here how such rules relate to the municipal law of Ghana. In that vein, Ghana is a dualist state as far as the relationship between municipal law and international law is concerned. Dualist states in international law theory are required to domesticate a ratified treaty before it becomes part of the national law. Mere ratification does not constitute a sufficient legal ground for a direct application of a treaty in the domestic legal system. An explicit act of incorporation is required. Again, national laws in the domestic legal system of dualist states take primacy over international law in the event of a conflict between the two. That is, in the domestic hierarchy of legal norms, national laws are at the apex. Any inconsistency that exists between the constitution of a dualist state and an international human rights treaty or law is thus expected to be resolved in favour of the former.

Article 75 of the Constitution is very clear that the execution of treaties, agreements, or conventions in the name of Ghana by the President is subject to an explicit ratification by an Act or resolution of Parliament. Human rights treaties that fall within this category are thus required to be domesticated before their enforcement in the domestic legal system by the national courts. However, the Supreme Court of Ghana seems to take a cautious view that domestication is not a necessary prerequisite in all instances as far as the reach and application of provisions of international human rights treaties are concerned. In *New Patriotic Party v Attorney-General [CIBA CASE]*[1996–97] SCGLR 729, the court held that it may not decline an invitation to apply international human rights instruments in the enforcement of the provisions of the Constitution merely on the grounds that such instruments are not incorporated into domestic law. Supreme Court Justice Atuguba reasoned in this case that the principles of international human rights instruments relating to fundamental human rights are enforceable to the extent that they fit into the provision set out in article 33(5) (p. 788).

Legal pluralism

Aside from the potential benefit of international law and, in this case, international human rights treaties as sources of law in Ghana, the general constitutional environment of the Ghanaian state also allows the application of the principle of legal pluralism. This is, we may note, of an internal nature as it has more to do with laws or legal orders from within as opposed to legal rules or orders from without. Legal pluralism, it is suggested, entails the existence of more than one 'legal order', according to which human 'behaviour' is regulated (Griffiths, 1986, p.1). This view is usually contrasted with 'legal centralism', which posits that law is uniform for all persons, exclusive of all other laws, and administered by a single set of state institutions (Griffiths, 1986, p. 3). In any case, legal pluralism could be conceived of as the co-existence of statutory law on one hand and customary law on the other.

Ghana is a culturally diverse society with not only numerous ethnic groups with varied cultural practices and norms, but also more than 50 distinct local languages thriving in hundreds of dialects (Atupare, 2014). The multiplicity of customary codes is a reality in Ghana. Thus, customary rules and practices of the various ethnic communities in Ghana form part of the laws of Ghana (Bimpong-Buta, 1983, pp. 132–132). The Constitution recognizes in article 11 the customary laws of the various ethnic groups as part of the corpus of the law in Ghana. Matters such as acquisition of rights in land, and family relationships – husband and wife, child and parent, marriage and divorce, and chieftaincy – that represent the core living embodiment of the country's cultural heritage, are all mostly governed by customary rules and practices.

To be more precise on this point, the Constitution provides in article 11 that the 'common law of Ghana' includes not only the rules of law generally known as the common law, but also 'rules of customary law', which are 'rules of law which by custom are applicable to particular communities'. In the light of this, the common law of Ghana embraces not only the general common law but also particular local customary norms (Bimpong-Buta, 1983). Therefore, rules of customary law are valid and enforceable if properly so held as rules of law, which are by custom applicable in a particular community (Atupare, 2014). The application of customary law in Ghana is further widened by article 26 of the Constitution that entitles a person to enjoy, practice, profess, maintain, and promote any culture, language, tradition, or religion. More so, article 39(2) enjoins the state to ensure that appropriate customary and cultural values are adapted and developed as an integral part of the growing needs of the society as a whole except for practices that are injurious to the physical and mental well-being of persons.

The constitution, the right to food, and gender equality in Ghana

With the constitutional environment of the state of Ghana determined, we are now on the proper pedestal to look into the quality of the Constitution on human rights, especially on the rights to food and gender equality. What is the position of the Constitution as far as the rights to food and gender equality are concerned?

The right to food as a fundamental right

It is important to state from the outset that under Chapter 5 of the 1992 Constitution, there is no specific provision on the right to food. But arguments to that effect can be inferred from the content of other provisions in the Constitution. A fundamental starting point is article 33(5), which can be construed to incorporate or recognize the right to food in the Constitution. This provision provides that rights that are considered to be inherent in a democracy and intended to secure the freedom and dignity of man, but are not specifically mentioned in Chapter 5 of the Constitution, shall not be regarded as excluded. So, rights such as the right to food would not be excluded from legal protection in Ghana because there is no express language in the Constitution providing for them.

Conceived as such, article 33(5) would seem to accord potential rights claimants, especially on the right to food, an opportunity to rely on its seemingly liberal words for a remedy in a court of competent jurisdiction. That is, the right to food, which is not expressly stated in the Constitution, can be read in through this provision.

Additionally, this right can be inferred from article 15(1), which provides that the dignity of all persons shall be inviolable. The many facets of human dignity will reasonably entail the right to food. That is, implicit in this constitutional protection of the inviolability of the dignity of all persons is a protection of aspects of the right to food (Vidar, 2006). The Constitution also protects the right to life in article 13. This right in human rights jurisprudence is seen as the supreme right from which no derogation is permitted even in time of public emergency. Even those not subscribing to a hierarchy of rights argument, where the right to life is at the apex, generally still consider the right to life as one of pre-eminence to which violations can never be remedied (Smith, 2003, p.194). This would seem to be the legal protection of the sanctity of life; the right to food appears to be central to achieving such a legal value.

The right to food as a right under the directive principles of state policy

Apart from this limited set of discrete provisions with relative connection to the right to food, the Constitution of Ghana also contains other more extensive provisions in relation to socio-economic rights listed in Chapter 6. These are known as the directive principles of state policy.

Inspiration could be drawn, for instance, from article 36(1) of the Constitution, which requires the state to take all necessary action to ensure that the national economy is managed in such a way as to maximize the rate of economic development and to secure the maximum welfare, freedom, and happiness of every person and to provide adequate means of livelihood and suitable employment and public assistance to the needy. This is supported by article 37(6)(b) that requires the state to provide social assistance to the aged such as will enable them to maintain a decent standard of living. We may therefore argue the point that the combined constitutional obligation of the state to secure the maximum welfare, freedom, and happiness of every person, provide adequate means of livelihood and suitable employment and public assistance to the needy and to ensure a decent standard of living for the aged would amount to an implicit recognition of the right to food in Ghana.

From the survey of constitutional provisions in respect of the right to food, either as a fundamental right or a right under the directive principles of state policy, it is obvious that the Constitution has recorded a low mark on direct express provision to that effect. That is, the Constitution of Ghana does not have a single provision that expressly and directly in a free-standing style provides for the right to food. Given the fact that Ghana is also a dualist country, it does not appear reasonable to expect direct reliance on or benefits from international human rights treaties in the domestic courts. At best, the indirect window of articles 13, 15, 33(5), 36(1), and 37(6) (b) is the most plausible option, coupled with a liberal judicial temperament to carry that through.

Gender equality

Unlike the right to food, gender equality is directly protected in Ghana under the Constitution. Article 12 makes all rights applicable to all persons without regard to any gender shades and article 17 makes all persons equal before the law. Article 17(2) states that 'a person shall not be discriminated against on grounds of gender, race, colour, ethnic origin, religion, creed, social or economic status'. However, article 17(4) makes provision for special legislation or policies to address discriminatory socio-cultural, economic, or educational imbalances in the Ghanaian society, while article 18 guarantees every citizen the right to property. Article 22 protects equal access of spouses to property jointly acquired during marriage, and assets that are jointly acquired during marriage shall be distributed equitably between the spouses upon dissolution of the marriage. Article 22 further prohibits the deprivation of a reasonable provision of a spouse's estate upon death or dissolution of marriage.

Article 27(1) of the Constitution mandates that special care shall be accorded to mothers during a reasonable period before and after childbirth, and during those periods, working mothers shall be accorded paid leave. Facilities are to be provided for the care of children below school-going age to enable women, who have the traditional care for children, to realize their full potential. Article 27(3) also states that women shall be guaranteed equal rights to training and promotion without any impediments from any person. Article 35 mandates the state to take appropriate measures to achieve reasonable regional and gender balance in the recruitment and appointment to public offices.

At the level of statute law, the Intestate Succession Law, 1985 (PNDCL 111), aims at providing a uniform intestate succession law applicable throughout the country regardless of gender. The Borrowers and Lenders Act, 2008

(Act 773) prohibits lenders from discriminating against a person on the grounds of race, gender, ethnicity, political affiliation, or religion. Section 3 of the Children's Act, 1998 (Act 560) makes it unlawful for a person to discriminate against a child on the ground of gender. The Domestic Violence Act, 2007 (Act 732) provides against domestic violence and gives a gender-neutral definition and target of domestic violence. Section 2 of the Legal Aid Scheme Act, 1997 (Act 542) entitles any person regardless of gender, for the purposes of enforcing any provision of the Constitution, to legal aid in connection with any proceedings relating to the Constitution if he or she has reasonable grounds for taking, defending, prosecuting, or being a party to the proceedings. More recently, the Lands Act, 2020 (Act 1036) in section 38 grants equal spousal rights to property acquired during the currency of marriage unless there is a written agreement to the contrary.

The courts, the right to food and gender equality in Ghana

The courts and the right to food as a fundamental right

Generally, the courts of Ghana are doing modestly well with human rights litigation. There is even a specialized High Court – Human Rights Division – for human rights claimants. However, there is a paucity of case law where claimants have approached the courts for the specific enforcement of the right to food. Perhaps, this is due to the absence of a specific clear constitutional provision and article 33(5) is rarely utilized. Nevertheless, there are some interesting pronouncements from the courts in respect of analogous economic, social, and cultural rights that could be used as a basis for an argument about the legal obligations connected to the right to food.

The first important case to consider is CIBA Case, where the court held, per Atuguba JSC, that it may not decline an invitation to apply international human rights instruments in the enforcement of the provisions of the Constitution merely on the grounds that such instruments are not incorporated into domestic law. Justice Atuguba reasoned in this case that the principles of international human rights instruments relating to fundamental human rights are enforceable to the extent that they fit into the provision set out in article 33(5).

By this reasoning, we have seen a singular judicial preparedness to activate article 33(5) in aid of human rights enforcement in the country. As indicated, article 33(5) provides for the inclusion of all manner of rights that are not expressly provided for in the Constitution. The right to food is one of such rights that is not provided for under Chapter 5. But this right exists in comparative constitutional democracies such as South Africa and does exist in international human rights treaties. It is thus a welcoming judicial posture as seen in the pronouncements of Justice Atuguba that article 33(5) might be deployed by the right to food claimants in the court of law.

Another important case to consider here is *Centre for Public Interest Law v Tema Oil Refinery (Suit No. E12/91/07 (2007.09.20 Unreported).* The first plaintiff is a non-governmental organization registered under the laws of Ghana and primarily engaged in the promotion of human rights and public interest litigation. The second plaintiff is an indigene and resident of Tema Manhean in the Tema Municipality. The plaintiffs jointly issued a writ against the defendant, a limited liability company set up and owned solely by the government of Ghana to refine crude oil into petroleum products for national consumption. The reliefs sought were, among others, a declaration that the defendant was negligent in spilling oil into the Chemu Lagoon, and very importantly for our purposes, a declaration that the oil spillage into the Chemu Lagoon was a violation of the rights of the inhabitants of Chemu, particularly the rights of those who were settled along the banks of the lagoon, to a clean and healthy environment under the Constitution and international law.

The reliefs sought did not, we must note, say anything explicitly about the right to food. It was the relief on protecting the plaintiffs' right to a clean and healthy environment and the court's judgment on same that brought us very close to issues of the right to food. The court found that the oil spillage led to the death of flora and fauna in the lagoon. This had affected the daily returns of the fisher-folk living around the lagoon. As in the contention of the plaintiffs and confirmed by the court, the fisher-folk became destitute. It was established that fishing is a source of both income and food for the inhabitants. So, a high level of pollution by oil spillage, as recorded in the case, would affect the people. Here, the court did not agree with a narrow conception of the consequences of environmental pollution. For the court, polluting the environment, as in water bodies in the instant case, could pulverize the capacity of people to access food.

The resistance of the court in this case to environmental pollution by a state institution is crucial in a number of ways. First, if the environment (the lagoon) is polluted, access to clean drinking water may be affected. This need not be for the people but domestic animals that people depend on as a source of food. Lastly, fishing activity that is a source of income and fish (meat/food) for the people would be severely affected. The overall effect of these consequences arising from environmental pollution is an avoidable destruction of the value of the right to food. It is reasonable to suggest that the corollary effect of protecting the right to a clean and healthy environment in this case is a judicial protection of the rights to food, water, and health.

Be this as it may, the courts of Ghana have also considered questions from some fundamental human rights provisions bordering on the application of international law with some amount of relevance to the enforcement of the right to food. For instance, in *Ghana Lotto Operators Association v National Lottery Authority* 2007–2008] 2 SCGLR 1088 (hereinafter referred to as the 'National Lotteries Case') at 1097, the court, citing article 6 of the

International Covenant on Economic, Social and Cultural Rights, 1966, acknowledged that the right to work is a human right that international human right instruments recognize. The court said it was prepared to help a human rights claimant if such a person were able to show with any clarity the extent to which such a right should be imported into Ghanaian law. By this holding, the court seemed to suggest an indirect protection of the right to food by analogous reasoning.

With these cases as reviewed above, we may say that the courts do not have any substantial bite in holding the right to food as a fundamental right. The limited decisions as considered show a lack of litigation in respect of the right to food and a lack of evidence to pronounce on the quality of the courts' performance on the right to food as a fundamental right. Notwithstanding this, it is reasonable to note that the courts have adopted the correct human rights posture with the material effect of protecting the right to food as a fundamental right. By their analogous reasoning principle, there is a vision from the courts to protect the right to food where such a right is entailed in another right.

The courts and the right to food under the directive principles of state policy

The right to food may also be protected as a directive principle as captured under Chapter 6 of the Constitution. Unlike the preceding section, this section examines the courts consideration of the right to food as a right under the directive principles. We begin with the observation that there are three sets of opinion in the prevailing jurisprudence of African countries as to the justiciability of the directive principles of state policy as rights. These sets of distinct judicial opinions can be represented by what I will term the *Absolute NO thesis*, *Relative NO thesis*, and *Presumptive YES* arguments.

Absolute NO thesis

In this group, the courts have taken the position that the only constitutional rationale for their inclusion in the Constitution is for the directive principles to serve as a source of guidance to the three arms of government in the discharge of their respective constitutional obligations. In that case, they are not at all justiciable in any court of law. For instance, in Nigeria, the Lagos State High Court held in *Morebishe v Lagos State House of Assembly* [2000] 3 WRN 134 that the directive principles in the Federal Constitution are not justiciable. Similarly in Lesotho, the Court of Appeal held in *Khathang Tema Baitsokoli and Another v Maseru City Council and Others* [2004] AHRLR 195 (LeCA 2004) that the directive principles cannot be enforced through the right to life as the former is not justiciable. It could be said that the position in these cases is that directive principles are declarations that cannot be independently enforced in a court of law as rights.

Relative NO thesis

The 'relative NO thesis' argument acknowledges the relevance of the fact that the principles are to guide the arms of government in the discharge of the responsibilities but suggests that they are justiciable when they can be properly tied to any of the justiciable fundamental rights stated in the Constitution. This used to be the position of the Supreme Court of Ghana but was abandoned. In *New Patriotic Party v Attorney-General* (CIBA Case, pp.743–744), Justice Bamford-Addo observed that while the directive principles cannot be enforceable in themselves, they are justiciable when read and applied in conjunction with any substantive guaranteed human rights and freedoms set out in Chapter 5. Her opinion was concurred with by the other Justices. From this, the court seems to take the position that, in general, while the directive principles are not in and of themselves legally enforceable by any court, there are particular instances where some of these provisions form an integral part of some of the enforceable rights in the Constitution. In such instances, they are justiciable of themselves because either they qualify as enforceable rights or can be held to be rights in themselves.

The presumptive YES thesis

This argument is to the effect that the directive principles are justiciable and rights under such principles will be justiciable and enforceable as a matter of free-standing rights. Unlike the 'relative NO thesis', which makes justiciability dependent on reading directive principles together with other rights, this argument does not depend on such joint reading of human rights provisions. In fact, the Supreme Court of Ghana in the *Ghana National Lotteries Case* takes the view that there is a presumption of justiciability of the whole Constitution, which includes the directive principles. The Supreme Court in this case revisited the question of whether or not the directive principles, which implicitly include the right to food, are justiciable. The response of the court to this question was that the entire Constitution, including the directive principles, is justiciable. The court reasoned that neither the Constitution nor its framers authorized that the directive principles are not justiciable.

Perhaps this test is not strange in human rights litigation as it affirms the view that no right is absolute. Enforcing the directive principles in Ghana is not merely a matter of conscience for successive governments, and that, unless the presumption is rebutted in relation to a specific directive principle, a piece of legislation or policy that violates the general or specific intent of the directive principles could be struck down by the court. In effect, the Supreme Court in the *National Lotteries Case* raised the justiciability of the directive principles to the status of a 'rebuttable presumption' in law. With this judgment in the system, we may say that the courts of Ghana have a positive record in aiding the protection of the right to food as a right under the directive principles.

The courts and gender equality

We must observe that the courts of Ghana did not start well with gender equality, especially when it relates to the application of customary law and women's rights. For instance, it was held in in *Mensah v Berkoe* [1975] 2 GLR 357 that even though a woman can acquire and hold property at customary law such property is said to belong to her family. Again, the court held in *Quartey v Martey* [1959] GLR 377 at p. 381 that where a woman assists her husband to acquire property or acquires it jointly with the husband, she has no proprietary interest in such property at customary law. By customary law, according to the court, it is the domestic responsibility of a man's wife and children to assist him in the carrying out of the duties of his station in life, e.g. farming or business. On that account, the court refused a woman's request for a declaration in *Gyamaah v Buor* [1962] 1 GLR 196 that she was entitled to a share of the 11 farms cultivated by her husband to which the defendant had succeeded. The court held that in Akan custom the woman was not entitled to a share of her late husband's property only on the ground that she assisted in the cultivation of the farms.

But it appears that the above position is different where assistance by a wife had been substantial and of a financial nature or there was an intention to share such a property. Affirming the rule, the court held in *Etua v Sika-Kyere-Abusua* [1982–83] GLR 781 that unless there is a clear intention to acquire a joint property, any property acquired by their joint effort is for the man solely. Again, it was held in *Abebreseh v Kaah* [1976] 2 GLR 46 that where a wife was found to have contributed substantially to the acquisition of the property, she would be given an interest in the property. This seems to be a clarification of the dictum of Hayfron-Benjamin J. (as he then was) in *Yeboah v. Yeboah* [1974] 2 G.L.R. 114 at p. 121 that customary law does not admit joint ownership by persons who are not related by blood.

But in a quite palpably discriminatory retreat, the courts have suggested that the doctrine of advancement applies where a husband purchases property in the name of a wife but does not apply where a wife purchases property in the name of a husband. In *Quist v George* [1974] GLR 1, it was held as well-settled that where a husband transferred property to his wife, the presumption of advancement was applicable. Therefore, the onus of proving that no gift was intended would be on the husband. However, no such presumption would arise when a wife transferred property in the name of her husband.

This position of the law prevailed until the courts seriously reviewed it by holding in *Mensah v. Mensah* [1998–1999] SCGLR 350 and subsequently in *Boafo v. Boafo* [2005–2006] SCGLR 705 that the principle of 'equality is equity' is the preferred principle to be applied in the sharing of joint property, unless in the circumstances of a particular case, the equities of the case would demand otherwise. In upholding the equality principle, the Supreme Court in *Nartey v Gati [2010]* SCGLR 745, speaking through Date-Bah JSC stated thus in respect of article 17 of the 1992 Constitution:

> Interpretation of Article 17
>
> This reference presents a genuine issue for interpretation because the concept of equality embodied in article 17 is by no means self-evident. To our mind, it is clear what article 17 does not mean. It certainly does not mean that every person within the Ghanaian jurisdiction has, or must have, exactly the same rights as all other persons in the jurisdiction. Such a position is simply not practicable. Soldiers, policemen, students and judges, for instance, have certain rights that other persons do not have. The fact that they have such rights does not mean that they are in breach of article 17. The crucial issue is whether the differentiation in their rights is justifiable, by reference to an object that is sought to be served by a particular statute, constitutional provision or some other rule of law.

With distribution of property between spouses upon the dissolution of marriage, the Supreme Court has held in the case of *Mensah v Mensah* [2012] 1 SCGLR 391 that a spouse may have an equal share or interest in the property acquired by the other spouse during the subsistence of the marriage. The court reasoned on pages 401–402 that:

> common sense, and principles of general fundamental human rights requires that a person who is married to another, and performs various household chores for the other partner like keeping the home, washing and keeping the laundry generally clean, cooking and taking care of the partner's catering needs as well as those of visitors, raising up of the children in a congenial atmosphere and generally supervising the home such that the other partner, has a free hand to engage in economic activities must not be discriminated against in the distribution of properties acquired during the marriage when the marriage is dissolved.

Nonetheless, the Supreme Court cautioned in the case of *Quartson v Quartson* (J4/8/ 2012) [2012] GHASC 49 (31 October 2012) that:

> The Supreme Court's previous decision in *Mensah v Mensah* …, is not to be taken as a blanket ruling that affords spouses unwarranted access to property when it is clear on the evidence that they are not so entitled. Its application and effect will continue to be shaped and defined to cater for the specifics of each case.

In the more recent case of *Fynn v Fynn and Others* (J4/28/2013) [2014] GHACA 129 (12 February 2014), the Supreme Court held that marriage is not a bar to an individual's capacity to acquire property alone. This is consistent, in the court's view, with article 18 of the Constitution, which gives individuals the legal capacity to acquire property either alone or jointly with others. Properties therefore acquired during the currency of marriage would not necessarily lead to joint acquisition and ownership and, under customary law, would not take away the rights of women to acquire and own such a property.

Meanwhile, the High Court (Human Rights Division) recently in *Commissioner, CHRAJ & 2 Ors Vrs Ghana National Fire Service & Attorney-General*, Suit No. HR 0063/2017-23-04 2018 [declared that regulation 33(6) of the Conditions of Services of the Ghana National Fire Service (GNFS) that required female employees of the GNFS to defer pregnancies until three years after their employment to avoid dismissal, is discriminatory and in effect, unjustifiable, illegitimate, and illegal. The court thus ordered the second and third applicants, who had been dismissed from the GNFS for becoming pregnant within the first three years of their employment and were thus dismissed under regulation 33(6) of the Conditions of Service, be reinstated into the employment of GNFS without prejudice to any benefits that accrued to them during the period of dismissal.

Selected evidence from the research field

As part of the study, various acts were found, which might be inconsistent with the basic tenets of the rights to food and gender equality. These include land grabbing; the practice of sacking farmers in order to create room for corporate mining; lack of markets for farm produce, which affects farmers' income-earning power and capacity to access food on the market; pollution of water bodies, which negates fishermen's efforts to generate income and to obtain fish as food; discriminatory cultural practices in relation to the acquisition of farm lands that undermine women; local taboos prohibiting the production of certain food crops by women; tension between Fulani cattle farmers and crop farmers over grazing lands; the Commission on Human Rights and Administrative Justice's determination of maintenance allowance as a livelihood remedy for women; tension between hunters or gamers and crop farmers with the practice of bush burning, thereby calling on district courts to settle agricultural related disputes; production of food at the family level with gender dimensions, and cultural farming practices that subordinate women.

From these generalities of evidence, there were some identified differences, especially in the East Gonja district, in the procedure of land acquisition between men and women. Respondents argued that, although both men and women can access land, women's land access is based on the purpose for which the land is acquired as well as whether the woman in question is an indigene or a migrant. For instance, one respondent remarked as follows:

> When it comes to real estate, women can own land without the knowledge of their husbands; even here in Salaga, women own more houses than men but when it comes to agricultural purposes, they cannot own land or inherit land without the knowledge of their husbands.
>
> (AA, male, Salaga, 17/03/2016)

That is, there are no direct customary rights to access such lands for farming purposes without routing it through the husband or family member. It was argued by the respondents that the principal reason for this gendered practice lay in the long-practised customary law that removed land from the set of rights that women can inherit. While this is supported by the general patriarchal nature of the communities, a respondent argued that

> men fought for this land years ago, not women, and on that account the direct beneficiaries can only be those who have fought for it. Women can only have an indirect access through these customary right holders.
> (SS, male, Salaga, 21/03/2016)

This is problematic for women as it is an unnecessary stricture for the acquisition of lands for agricultural activities. But these discriminatory rules applicable to women in the acquisition of land do not apply to titled women. A respondent stated that for a woman to get access to land, it depends on the type of woman in question (DS, male, Salaga, 21, /03/2016).

But we should note that there was a minority view from the Gonja East district to the effect that customary rules of inheritance do accept women's right to inherit land from their fathers. Be this as it may be, there are notable gender differentials and preferences in the labour practice and reward. For instance, women are generally preferred to men for the sowing or planting and harvesting of all crops except yam. On the other hand, men are preferred in weeding and raising of mounds for yam cultivation. Depending on the time and farming activity in question, the gendered labour practice sometimes creates disabilities for earning capacities of men and women, and this sometimes affects their capacity to generate income and access food on the market.

Besides, some of the farmers often take loans from banks and microfinance companies such as Sinapi Aba Savings and Loans. The monies are used in buying fertilizers, chemicals (such as weedicides and pesticides), or seeds. The difficulty is in the collateral demanded by the lending institutions. This is sometimes a registered property such as land or a house of a certain quantified monetary value. Individual farmers are not able to meet this traditional banking requirement, especially women at the local level who cannot own land. To deal with this, it was reported that women mostly organized themselves into groups, a practice favourably looked at by the lending financial institutions, which developed a special scheme for the purposes of granting them loans for their farms. Though this practice lowers the bar for the women to be able to access the loans, it is generally unsatisfactory. The money usually given has to be shared in equal terms among the women in the group. This results in mere fractions of the money going to individual women.

The traditional food production system at the family level was gendered. While men are responsible for the provision of yam and maize, which are the main staples, women provide vegetables, spices, and any other things needed for soup preparation. This gendered approach to food production in the family has placed an additional burden on women in this area. This is because the difficulties with the availability of the staple foods such as yam and maize are not as troubling as those pertaining to ingredients. Women cannot sell yam or maize for cash to be able to buy spices or ingredients. The men generally frown upon this, and women remain helpless and sometimes suffer some form of domestic violence from this predicament.

It was observed that some women, especially in Grushie Zongo and Salaga in the East Gonja district, have to engage in barter trading on the banks of the Mankango River where fish is exchanged for yam or maize or cassava. They might have to do this undercover away from male family members or husbands. But the success of a woman with this on a particular day depends on the fortunes of the fishermen in catching fish and the correlative demand for yam or other food items. This was mostly tasking as some women spent the whole day on the banks of the river trying to get fish for food. It was reported that this practice was a source of emotional problems for women, particularly those whose labour had already gone into the production of the yam on family farms that the man had used to fulfil his side of the obligation in family food provision.

Lastly, the pollution of rivers and lagoons through mining and refinery activities presents a threat to gender equality and the right to food. Those who depend on the rivers and lagoons for fishing activities have lost food (fish), income (sale of fish), water, and employment. Their economic power earned through the sale of fish is destroyed, and, therefore, they cannot access food on the market even if such food is available; reasonably connected to this, they cannot even have the fish as part of the food to be consumed. So the adequacy of the food component is undermined. This point bears resemblance to the tension between cattle farmers and crop farmers. Various degrees of destruction are caused to crops by animals, such as cows, sheep, and goats, which are kept in large numbers. The competition is on grazing lands. This has generated tension between the right of the cattle farmer to keep his animals alive as part of the efforts to meet the requirements of the right to food and the crop farmer's right to be saved from destruction.

Conclusion and recommendations

To conclude, the human rights constituency in Ghana is not necessarily dulled by the poverty of direct express provisions in the Constitution on the right to food. As a matter of principle, the right to food could still be pressed and pursued by relying on a strong court. The courts have shown a remarkable courage and right judicial rectitude in the appreciation of the Constitution to the extent of recognizing the capacity of article 33(5) to deal with unenumerated rights such as the right to food in Ghana. In contrast, the gender equality provisions will not be an end in themselves. Though the courts of Ghana have shown a sufficient degree of the right attitude towards serious reforms on equality claims, it remains an aspiration that certain cultural practices that throw dust on such judicial efforts would be eschewed. Collective efforts at all levels would be crucial in order to give true meaning to the liberal provisions on gender equality in the Constitution of Ghana.

We should note that as a country emerging from military dictatorship, the potential of article 33(5) and express provisions on gender equality can only be realized if the bar is active enough on human rights litigation. Note that courts will not go to town to look for cases on human rights. This can only be done when applicants appear through counsel. It is our contention that the principles of human rights must be imbued as part of the core properties of law in the country. In a political community like Ghana, law means nothing for the suffering population, the aged, weak, sick, and malnourished, and above all those without food and water. The insurance against this is a denunciation of all manner of acts that disdain rights, particularly the right to food and gender equality.

Bibliography

Atupare, P.A. (2014) 'Sarbah on Legal Pluralism and Sense of Legality in the Gold Coast: Some Insights for Ghana's Current Legal System', *University of Ghana Law Journal* (XXVII) pp. 100–124

Bimpong-Buta, S.Y. (1983) 'Sources of Law in Ghana', *Review of Ghana Law* (15) p. 129.

Elazar, D. (1985) 'Constitution-Making: The Pre-eminently Political Act', in Keith G. Banting and Richard Simeon (eds) *Redesigning the State: The Politics of Constitutional Change in Industrial Nations*, Toronto, University of Toronto Press, pp. 232–248.

Griffiths, J. (1986) 'What Is Legal Pluralism', *Journal of Legal Pluralism and Unofficial Law* (24) p. 1.

Kornblith, M. (1991) 'The Politics of Constitution-Making: Constitutions and Democracy in Venezuela', *Journal of Latin American Studies* (23) pp. 61–89.

Riker, William H. and Peter C. Ordeshook, (1973) *An Introduction to Positive Political Theory*, Englewood Cliffs, N.J., Prentice-Hall.

Smith, R. (2003) *Textbook on International Human Rights*, Ashland, Oregon: Blackstone Press.

Vidar, M. (2006) State Recognition of the Right to Food at the National Level, Research Paper No. 2006/61, United Nations University-World Institute for Development Economics Research.

Wheare, K.C. (1966) *Modern Constitutions*, Oxford, England: Oxford University Press.

Decided Cases in Ghana

New Patriotic Party v Attorney-General [1996–97] SCGLR 729
Centre for Public Interest Law v Tema Oil Refinery Suit No. E12/91/07
Ghana Lotto Operators Association v National Lottery Authority [2007–2008] 2 SCGLR 1088
Mensah v Berkoe [1975] 2 GLR 357
Quartey v Martey [1959] GLR 377
Gyamaah v Buor [1962] 1 GLR 196

Etua v Sika-Kyere-Abusua [1982–83] GLR 781
Abebreseh v Kaah [1976] 2 GLR 46
Yeboah v Yeboah [1974] 2 GLR 114
Quist v George [1974] GLR 1
Mensah v Mensah [1998–99] SCGLR 350
Boafo v Boafo [2005–2006] SCGLR 705
Nartey v Gati [2010] SCGLR 745,
Mensah v Mensah [2012] 1 SCGLR 391
Quartson v Quartson (J4/8/ 2012) [2012] GHASC 49
Fynn v Fynn and Others (J4/28/2013) [2014] GHACA 129
CHRAJ & 2 Ors v Ghana National Fire Service & Attorney-General (Unreported judgment of the High Court, Accra, Human Rights Division, in Suit No. HR 0063/2017, 23 April 2018)

Decided Case in Lesotho

Khathang Tema Baitsokoli v Maseru City Council and Others [2004] AHRLR 195

Decided Case in Nigeria

Morebishe v Lagos State House of Assembly [2000] 3 WRN 134

From the unequal harvests of commercialization to the right to food and gender equality

What roles for governments, agribusinesses, and rural communities?

Joanna Bourke Martignoni, Christophe Gironde, Christophe Golay, Elisabeth Prügl, and Dzodzi Tsikata

How does the commercialization of land and agriculture impact food security? How do local, national, and international norms and policies shape these outcomes? And how do processes of commercialization and associated policies, laws, and practices affect gender equality and the right to food? These three inter-related questions guided the research conducted under the DEMETER project, and they also provide the structure for this book. By probing the impacts and politics of commercialization in Cambodia and Ghana, the chapters highlight not only complexities, but also similarities and common patterns. This concluding chapter provides a comparative analysis of our findings through the lenses of the right to food and gender equality. We explore outcomes and processes in the areas of food availability, accessibility, and adequacy and examine the implications of our findings for future policies and research.[1]

The realization of the human right to food implies that its core elements of availability, accessibility, and adequacy (which includes cultural acceptability), are respected, protected, and fulfilled. Food is a human right for everyone, and this means that particular attention must be paid to preventing and redressing inequalities within food systems. In the DEMETER project, we used a gender perspective to gauge the availability, accessibility, and adequacy of food in contexts of commercialization, bearing in mind that gender intersects with other dimensions of inequality, here in particular those related to class, age, and ethnicity.

Food availability

Agricultural development policies in both of our case countries emphasize interventions to increase the availability of food through commercial food production, and these interventions have borne fruit. Food availability has

DOI: 10.4324/9781003202004-16

increased in the aggregate and food supply tends to be more diverse. As the interviewees quoted in Gironde et al.'s chapter affirm for Cambodia, there is now more food available, it is available all year round, and it is perhaps more diverse ('it's better than before'). In her chapter, Reysoo shows that for Kampong Thom, food reaches even the most remote places through open markets, shops of various sizes, and mobile food sellers transporting their wares into villages. The picture is somewhat more nuanced in Ghana, where in spite of a long tradition of rural food markets and observed reductions in levels of child malnutrition, seasonal food shortages continue to be the norm in both northern and southern Ghana. Here, interviewees told us that staple foods become scarce during the lean season, in particular in areas where most productive land is dedicated to non-food cash crops, and women are faced with the emotional and practical burdens of figuring out how to feed their families. In their chapter, Dzanku and Tsikata diagnose the phenomenon as over-commercialization; that is, although poor households earn income from cash crops, without efficient food markets, specialization in cash and non-food crops makes food less available and more expensive. Indeed, average food security levels appeared lower in the Asunafo-North district, a highly commercialized area devoted to cocoa, than in the Gonja-East district, where farmers produced food crops in commercial quantities. Thus, despite an overall increase in agricultural production, the substitution of food crops with non-food cash crops failed to guarantee year-round food security. Functioning food markets and the devotion of some land to food crops emerged as two key prerequisites for securing food availability under conditions of commercialization.

The production of some of their own food staples by farmers continues to be important for ensuring food availability in both Cambodia and Ghana. Crucial to this is access to land, which is mediated by gender and ethnicity. Commercialization has reduced the accessibility of land for poorer people, including many women and indigenous peoples. This growing land scarcity alongside the enclosure of formerly communal forest lands has resulted in a decline in the availability of 'free' food sources and this has important repercussions for the realization of the right to food as well as for the burden of food provisioning, which is typically borne by women. In Ratanakiri, Cambodia, indigenous communities have lost large stretches of land because of economic land concessions and these households are particularly food insecure, as shown by Gironde et al. Loss of land has meant the loss of both homestead gardens and plots dedicated to growing staples. This is paired with the declining availability of food from forest sources and the loss of communal areas for livestock grazing. Similarly, in Asunafo-North, Ghana, in addition to households now having to buy staple foods in weekly open-air markets, resources harvested from the commons, such as leafy vegetables, cocoyam, mushrooms, snails, and bush meat have become scarce, and only 'the rich men are the ones who can afford' such foods, as reported by Eweh and Tsikata in their chapter.

Degrees of commercialization differ between female and male farmers. For Cambodia, the chapters by Gironde et al. and Bourke Martignoni and Joshi demonstrate the continued role played by women smallholder farmers in the production and sale of vegetable crops and small livestock. Male farmers in Cambodia are more likely to focus on investments in 'boom' cash crops for export such as cassava, rubber, and cashew. The same scenario was observed by Dzanku and Tsikata and by Eweh and Tsikata in Ghana with some important regional variations in that women in the more commercialized, matrilineal zone in the south were producing lucrative commercial crops at a similar rate to men. In both countries, the capacity of women smallholder farmers to engage in large-scale commercial commodity production was hampered by their tenure insecurity and comparatively smaller landholdings and patriarchal power relations that dictate that women should provide labour for family farming enterprises rather than commanding male labour for their own farms.

In Ghana, the chapter by Dzanku and Tsikata shows that female-headed households have much smaller landholdings and are less likely to hold land titles than men. In a context of long-term commercialization, land is increasingly acquired through the market rather than through lineage allocation and inheritance, often leading to a further loss of land rights for women. In the most highly commercialized study region in Ghana, Asunafo-North, only 58% of farmers interviewed had inherited their land, as compared with 94% in Garu Tempane in the less commercialized north. As share contracts and purchases become the predominant means of accessing land in southern Ghana, women's land rights, particularly those over lineage lands and the commons, have steadily eroded (Dzanku et al., 2021). In the indigenous communities of Cambodia, described by Beban and Bourke Martignoni, customary inheritance norms are in flux, with some families moving away from a matrilineal system that provided women with a greater degree of status to a bilateral pattern where small amounts of land are equally shared amongst all children in the family. But even when women have access to jointly titled land, Gironde at al. and Bourke Martignoni and Joshi show that they do not have the same control over it in that they do not have the power to decide how it will be used or when it might be sold.

Land loss entails the loss of a key social protection mechanism. As Beban and Bourke Martignoni explain, young people gain a start in agriculture through inheritance, while older people secure care and assistance with farming tasks by providing land to children who look after them. They argue that in Cambodia, rapid agricultural commercialization has created a gendered crisis of care, as growing land scarcity undermines informal mechanisms of social security in a context where there is limited state support. Commercialization thus functions to sharpen inequalities by fostering land concentration and landlessness, with women, younger, and older people as well as some indigenous populations being particularly affected.

Accessibility

The availability of food does not mean that it is accessible to everyone. As many of Reysoo's Cambodian respondents emphasized: 'Today you can buy anything at the market whenever you want to, provided you have money'. The statement touches upon the issue of purchasing power and the underlying physical, structural, and institutional barriers to accessing food. Not all individuals and families have equal capacities to buy food all year round. Thus, in addition to seasonal food shortages resulting from reduced food crop production among smallholders, and weak food markets, there are also seasonal shortages resulting from market fluctuations. In Cambodia, these shortages take two forms: fluctuations in the price of food items consumed, and fluctuations in rural job markets. In Ghana, the shortages are manifested in the lack of ready availability of preferred foodstuffs, and food price volatility.

In Cambodia, access to food has become a source of stress for many land-poor families, who have become entirely dependent on cash or credit to buy food. Demand for wage labourers, and hence access to cash, is seasonal and, as Gironde et al. show, seasonal food shortages correlate with seasonal variations in wage labour opportunities. These are predominantly in agriculture and consequently vary according to crop production cycles. Over time, seasonal food shortages have increased in Ratanakiri and decreased in Kampong Thom between 2016 and 2019 in parallel to the differences in wage employment and non-farm self-employment opportunities in the two provinces. In Ghana, commercialization in a context of expansion in education, rural stagnation, and poor terms and conditions of casual labour, has created massive labour shortages that are addressed through the extensive use of weedicides. Thus, the lack of wage labour opportunities is not seasonal, but instead is an integral part of the agrarian economy (Dzanku et al., 2021).

To cope with food shortages in Cambodia, people resort to borrowing money, leading to high levels of indebtedness and forced land sales. While out-migration from rural to urban areas in search of wage labour opportunities is observed in many agrarian communities in response to food insecurity, the chapter by Beban and Bourke Martignoni shows that this is not a livelihood strategy that is viewed as culturally acceptable or feasible for many indigenous women, given the structural discrimination in the labour market. Both the Tsikata and Eweh and the Gironde et al. chapters find that, given their responsibility for putting food on the table, women play a far more important role in borrowing money or purchasing food on credit than men do. They also found evidence of women in poorer households reducing the number of their meals and eating lower-quality food items during periods of food scarcity. Commercialization thus introduces new inequalities in access to food in both Cambodia and Ghana.

Commercialization also changes labour arrangements in ways that are profoundly gendered. This includes an increase in women's productive and reproductive labour burdens, changes in the gender division of labour, and the payment of unequal wages. The increase in women's labour burden is particularly marked in Ghana. Eweh and Tsikata show in their chapter that there is an expectation that women in rural Ghana should work on the male household head's plots, and as a result, commercialization has increased the demand for women's unpaid labour. At the same time, as explained in the chapter by Atupare, customary expectations that women contribute fish, vegetables, and other ingredients to the family diet have not changed, leading to a double or triple burden for women as the major providers of household nutrition (see also Atupare, 2016). Conversely, women farmers are unable to access the labour or the land they need to participate independently in contract farming schemes.

These patterns differ in Cambodia, where we found labour to be more commercialized. As has been described in other contexts, moving work out of homes and farms leads to a physical separation of income-earning and reproductive labour, what Joshi (2020a), following Maria Mies, calls women's 'housewifization'. Unpaid reproductive labour remains women's work, hindering their participation in the labour market. Bourke Martignoni and Joshi show that women wage labourers in Ratanakiri, Cambodia regretted that they could not bring their children to commercial plantation workplaces, and if they did so, they were paid less (see also Joshi, 2020a). Not surprisingly, Gironde et al. and Beban and Bourke Martignoni confirm that women farmers in Ratanakiri – particularly those who are divorced or older – are more likely to engage in reciprocal labour exchanges, while men prefer to use paid agricultural labourers.

The sharpening gender division of labour associated with commercialization comes with unequal wages and discriminatory agricultural labour markets. In highly commercialized areas of Cambodia, Gironde et al. found that job creation favoured men, and this meant they were much more likely to participate in wage labour than women. Factors such as greater physical strength and the capacity to use heavy tools, mobility, and the greater ease for men to work at night explain employers' preference for men over women and are frequently used to justify the gender differences in wages. In the northeast of Cambodia, most jobs were taken by immigrant Khmers, leaving indigenous and local populations marginalized (Joshi 2020a). Studies of plantation and contract farming agriculture in Ghana have found a similar pattern of a persistent division of labour. Men are typically assigned jobs considered to rely on physical strength, while women are employed in labour-intensive tasks that play to gender stereotypes about their greater attention to detail and manual dexterity. These differences have been used to justify consistently lower wages for women in agriculture across Ghana. Furthermore, contract farmers and sharecroppers are also more likely to be men than women (Tsikata and Yaro, 2014).[2]

The inequalities in labour arrangements generated by land and agricultural commercialization have implications for food security. The greater food insecurity of indigenous populations in Cambodia has already been mentioned. Gender differences, in contrast, show a more complex picture. Women in Cambodia reported that they had eaten smaller or fewer meals and gone hungry at night at a higher rate than male respondents. Although our surveys show that the percentage of Cambodian women having to cut food intake decreased between 2016 and 2019, this gender difference has persisted. In contrast, in Ghana, while women and men eat differently in terms of quantity and quality of food, the survey data did not show gender differences in terms of hunger. Access to income matters, but clearly it is not the only determinant of individual food security. Although there is increased purchasing of prepared food in contexts of commercialization, most food continues to be cooked by women and consumed in households. And the gender-differentiated access to and quality of food consumed is often mediated by food cultures.

Adequacy and food cultures

An often-overlooked element of the human right to food is the notion of food adequacy, which encompasses dietary diversity, food safety and quality, and access to culturally appropriate food. Our conception of cultural appropriateness in this book takes into account the availability and consumption of novel food products which become part of a people's food culture. In Ratanakiri, for instance, ideas about the cultural adequacy of food within indigenous communities are undergoing a transformation. The introduction of new food products, higher incomes, the decreased availability of wild foods and the adoption of Khmer-style cooking are all influencing the types of foods that are sourced and prepared. Commercialization has also been accompanied by a surge in food-related businesses as well as increased possibilities for men to eat outside the home (Reysoo, this volume). Similarly in Ghana, north–south migration, the availability of new varieties of imported and locally produced food in food markets across the country, and the increasing consumption of purchased meals, have expanded notions of culturally acceptable food in our research communities (Eweh and Tsikata, this volume). In discussions with our interlocutors, it becomes clear that cultural adequacy intersects with financial accessibility, as many people make trade-offs between what they can afford and what they find adequate. Cultural adequacy must, therefore, be approached as a dynamic and understudied element of the right to food that is appreciated in different ways by individual right holders depending upon their own preferences and expectations about what makes for adequate food.

In Cambodia and Ghana, women tend to have responsibility both for sourcing food and for cooking. As a result, they play a key role in assuring nutritional adequacy in a context of commercialization, and they do so within the confines of gendered customs, while also subverting these in their practices. In both Ghana and Cambodia, what is considered adequate food is changing and often today depends on what is affordable and what is convenient. Broadly speaking, in both countries, people tend to consider food as adequate if they can access staple foods, a construction that is profoundly gendered. Eweh and Tsikata show that in northern Ghana, staples (in particular *tuo-zaafi*, prepared from maize or millet) are considered more nutritious and are valued more highly than proteins, in part because they constitute men's contribution to the household diet. In the south of Ghana, people associate food shortages with a shortage of *fufu* (pounded cassava or plantain), which is the preferred staple in the region. Similarly, in Cambodia, rice is the central part of a meal with fish, vegetables, and sauce regarded as optional accompaniments added when available or affordable. However, although food security is associated with eating enough carbohydrates, added protein is highly valued in the study sites in both Ghana and Cambodia.

Gendered food cultures also affect what, where, and how much food is eaten. In both of our case countries, men tended to have richer diets. As shown in the chapter by Eweh and Tsikata, in Ghana, men typically eat separately from the rest of the family and consume more protein. They also are more likely to have opportunities to eat outside the home, supplementing their food intake with various forms of snacking. Similarly, in Cambodia, meat is considered an energy-rich food and, when money is scarce, it is reserved for the working men of the family. As in Ghana, Cambodian men were also more likely to eat meals outside the home than women, and they were far more likely to consume alcohol. In Kampong Thom, Reysoo reports from the DEMETER survey a slightly higher dietary diversity score for men than for women. For Cambodia, more broadly, Gironde et al. find that women's responsibility for putting food on the table often leads them to sacrifice and eat less and lower-quality food.

Our research revealed old and new practices of food consumption in rural communities arising from the changes induced by agricultural and land commercialization. According to Eweh and Tsikata, in the colonial period, migrants from other parts of southern Ghana to the cocoa-growing frontier areas brought with them a new way of processing low-quality cassava so that it could be stored and consumed during the lean season. More recently, maize-based foods, yam, and rice have become staples all over Ghana, as has canned fish. On the other hand, commercialization in Ghana has reduced dietary diversity and food quality, according to some respondents. In southern Ghana, foods that used to be available in the wild have become scarce because of the intensive use of agro-chemicals. Similarly, in the upland areas of Cambodia, wild vegetables have become harder to source as a result of the enclosure of common forests, and indigenous households are adopting the more meat-heavy cuisine introduced by Khmer migrants. These transformations of eating and food cultures are not always regarded as negative, and some of our Cambodian respondents noted that they appreciated the convenience of being able to purchase 'modern' food from the market, with several women observing that the availability of purchased food means that they no longer have to get up early to cook food before leaving for work in the morning (Beban and Bourke Martignoni, 2021).

There is ambivalence about changes in the quality of available food. The villagers from Reysoo's study in Kampong Thom lamented the fact that market-bought vegetables and fish contain chemicals and are less fresh and less tasty; and they felt that food contamination was making them ill. In both Ghana and Cambodia, people also spoke of the introduction of unhealthy new convenience foods. In Ghana, this was associated with processed foods consumed mostly in wealthier households. In Cambodia, Reysoo indicates a strong presence of industrially produced snacks and sweets considered as modern and thus desirable but also nutritionally empty. Thus, even if commercialization may have made food more available and accessible, low-quality foods may generate health impacts, such as obesity, that undermine the realization of the right to adequate food and nutrition.

In sum, land and agriculture commercialization has fallen short of realizing the right to adequate food. As shown in the case of Ghana, the existence of efficient markets and certain levels of income cannot be assumed. In Cambodia, commercialization has deprived smallholder farmers of land without creating decent work opportunities or comprehensive social protection schemes to enable rural people to purchase food. Commercialization also has fallen short because the quality of food often deteriorates as the availability of wild food declines and snack food comes to substitute more nutrient-rich, home-grown food. And significantly, processes of commercialization reproduce and sharpen inequalities. These inequalities are apparent in the depletion of the gendered reproductive labour that enables household food security, the conversion of communal forest and food crop farmland to cash crop plantations, and growing landlessness amongst poor smallholder farmers, particularly women and younger people.

The role of policies and human rights guarantees

Commercialization has been propelled by neoliberal policies, but the international food regime also promotes human rights, and governments combine the drive to commercialization with other policy goals such as food security and human development (Graziano Da Silva, 2019). Both Cambodia and Ghana are bound by international human rights obligations related to the promotion and protection of gender equality and the right to food (Bourke Martignoni, 2021, 2018). The countries are parties to the International Covenant on Economic, Social and Cultural Rights and to the Convention on the Elimination of all Forms of Discrimination against Women (CEDAW), and both voted in favour of the adoption of the United Nations (UN) Declaration on the Rights of Peasants and Other People Working in Rural Areas in 2018 (Claeys and Edelman 2019). In Ghana, according to Atupare's analysis in this book, while the 1992 Constitution makes no explicit reference to the right to food, it directly protects gender equality. Nevertheless, courts in

Ghana have protected the right to food on the basis of directive principles of state policy, which are concerned with ensuring the welfare and livelihoods of citizens. In contrast, in Cambodia, the right to food and gender equality are both explicitly acknowledged in legislative and policy documents; however, the neo-patrimonial legal system means that appeals to judicial processes are usually either inaccessible or counter-productive and dangerous for most rural people (Joshi, 2020b; Bourke Martignoni, 2021).

Formal human rights guarantees founder in practice and fail to be 'vernacularised' or appropriated at the local level for a variety of reasons (Merry and Levitt, 2017). First, as the case of Cambodia illustrates, they often stand in tension with policies of commercialization. In their chapter, Bourke Martignoni and Joshi show how joint private land titling, a key neoliberal prescription, becomes a driver of large-scale land acquisitions and dispossession in rural Cambodia. In addition, the proliferation of debt through the spread of microfinance institutions leads to land loss. Commercialization policies thus seem to invite dispossession. Commercialization policies also tend to come with a commitment to a small state. As a result, they are rarely accompanied by state-provided social security. As Beban and Bourke Martignoni show for the rural upland communities in Cambodia, the state appears as a mediator of global capital and an enforcer of private property rights for those with the right political connections and economic power. But it is absent in terms of public social service provision, although the right to social security is guaranteed (for Khmer citizens) in the Cambodian constitution.

Existing political structures provide a second set of obstacles to implementing rights-based approaches. Joshi et al. illustrate the salience of patronage in the Cambodian 'hybrid state', which not only marginalizes gender equality policies but also violates procedural rights associated with the right to food. Male patronage networks limit the expertise and funding available to advance gender equality policies and the illiberal state has increasingly turned processes of consultation with civil society into an empty performance. While Ghana's political system is more open than Cambodia's, Awo and Antwi's chapter diagnoses shortcomings with regard to women's participation in agricultural policymaking, both at the national and sub-national level, which amount to no more than 'tokenism'. Not surprisingly, they argue, significant gender gaps persist in Ghanaian agriculture.

Informal norms are a third potent barrier, particularly with respect to the implementation of gender equality legislation and policies. Bourke Martignoni and Joshi illustrate this for the case of joint land titling: although government documents frequently highlight the success of joint-titling legislation, in practice, the vast majority of women in the country have not benefitted from more equitable access to land as a result of neoliberal cadastral reforms. Patriarchal norms stand in the way. These norms are also embedded in the state and slow the implementation of gender equality policies. Joshi et al. identify norms pertaining to property rights, labour expectations, and prescriptions for virtuous behaviour that hinder the implementation of gender mainstreaming in Cambodia. Nevertheless, as Torvikey and Atupare highlight in their chapter, settings of normative pluralism may potentially open up new opportunities and spaces for women to strategically engage in 'forum shopping' in order to claim their land and labour rights. In the same vein, Bourke Martignoni and Joshi show how women's resistance to land and resource-grabbing by corporate and state actors may lead to the redefinition of gendered subjectivities.

The Ghanaian case illustrates a fourth difficulty for right-to-food policy processes, namely the dependence on international donors and the salience of agribusiness corporations that neoliberal ideologies have invited into policy-making. While the government has designed several agricultural policies at both the national and sub-national levels geared towards attaining zero hunger (SDG-2) and gender equality (SDG -5), the implementation of these policies is fraught with fiscal challenges. Decentralization opens up avenues for private sector actors to fill the funding gaps. As a result, major agribusinesses drive the implementation of government policies at the sub-national level, co-opting key government agencies for agricultural policies that align with their interests. Specifically, they promote commercialization and cash crops at the expense of food crops. Interestingly, however, these actors also promote women's participation in agriculture, and they appear receptive to the incorporation of international gender equality norms and standards within their operations. In practice, these corporate social responsibility initiatives tend to adopt a neoliberal approach to inserting women within existing agricultural labour markets and contract farming schemes without meaningfully addressing the gendered power relations that perpetuate inequalities (Tsikata, 2014; Lanz et al., 2020).

In sum, the juxtaposition of policies of commercialization and human rights in the international food regime yields contradictory outcomes. The reification of individual property rights and the enclosure of the commons, the under-emphasis of social protection, and the privileging of the interests of financial and agribusiness actors often stand in stark tension with the promotion and protection of the right to food. In addition, male patronage networks, patriarchal gender norms, and a lack of meaningful consultation with relevant right holders, including women, are obstacles to integrating gender equality considerations into agriculture and food security policies. Commercialization policies that work for those at the margins of power would require addressing these obstacles and radically redesigning the governance of food systems.

Policy recommendations

Our research in Cambodia and Ghana shows the way in which commercialization limits the right to food:

- A focus on cash crops may produce shortages in food crops for smallholder farmers. It also severely restricts access to wild foods and forest products through logging, the privatization of the commons, and the use of agro-chemicals.

- Commercialization sharpens and creates inequalities based on gender, class, and ethnicity, reinforcing gender divisions of labour and discrimination in access to land, increasing women's burden of unpaid reproductive labour while providing limited income-earning opportunities. These gendered inequalities intersect with class and ethnicity, making access to food less secure for many rural people.
- The potential of commercialization to increase the variety of foods people have access to is moderated by its tendency to undermine existing, accessible food supplies, and by the low quality of the new foods. Food cultures are also deeply gendered, with men often eating greater quantities of food and having more diverse diets than women. The interaction between commercialization and food cultures can therefore deepen gender inequalities in access to food and nutrition.
- Governments and agribusinesses have legal responsibilities to respect and protect the rights to food and gender equality and to provide appropriate remedies in the event that these rights are violated. Human rights obligations are insufficiently acknowledged in contemporary agricultural policies, which promote commercialization and increased food production while neglecting the dimensions of food accessibility and adequacy.
- To avoid the harmful human rights impacts of commercialization, food security, land governance, and agricultural development policies and programmes should be grounded in participatory frameworks that prioritize the promotion and protection of economic, social, and cultural rights (Bourke Martignoni, 2020). The value in using feminist right-to-food methodologies that highlight the need for inclusive, democratic participation in food policymaking to challenge existing, patriarchal power structures should also be emphasized (Bourke Martignoni, 2021). States and agribusinesses should respect and protect the right to food in rural communities and promote gender equality. This is particularly urgent in the wake of the Covid-19 pandemic and in light of the impact of climate change, both of which have increased socio-economic inequalities (Bourke Martignoni, 2020; FIAN International, 2020a, 2020b, 2020c). The adoption of the UN Declaration on the Rights of Peasants and Other People Working in Rural Areas in 2018, which emphasizes the rights to food, land, decent employment, and social security, offers an additional tool to guide states' behaviour (Golay, 2019, 2020).

A number of concrete policy suggestions follow from our studies. To begin with, governments should promote *crop diversification*, instead of focusing on a few selected crops, to ensure the enjoyment of the right to adequate food. In Ghana, the state, along with donors and agribusinesses, prioritizes cash crop production, while efforts to meet food and nutrition needs through increased local production should occupy an important place in agricultural policy. Eweh and Tsikata suggest that crop diversification has the potential to prevent food shortages resulting from the poor performance of particular crops while expanding the nutritional content of the food that is available in rural communities.

There also needs to be a stronger policy focus on the elements of food accessibility and adequacy. Accessibility requires a firm focus on creating *decent jobs* in the countryside. It also requires governments to implement universal *social protection mechanisms*, as argued by Beban and Bourke Martignoni. In both Cambodia and Ghana, comprehensive social security coverage is inaccessible for the majority of rural people, particularly women, who are excluded from the formal employment relationships that would enable them to receive old age pensions, sickness and caring benefits, and other entitlements. With regard to food adequacy, it may be useful to integrate the perspectives not only of the rural people who consume food, but, as Reysoo argues, that of local food vendors, who in Cambodia offer a wide array of affordable, accessible food items, which are increasingly the nutritional foundation for low-income households. In the case of Ghana, Eweh and Tsikata argue that policy responses to food insecurity need to address more consistently the inequalities in the distribution of productive resources between men and women and the lack of recognition of women's critical role in household food provisioning.

Other recommendations focus on processes of policymaking and implementation. Ankrah et al. urge the Government of Ghana to provide realistic budgetary allocations through innovative local revenue mobilization at the sub-national level to reduce the high fiscal dependency on agribusinesses. They also recommended that agribusinesses should be encouraged to pay more attention to local institutions responsible for natural resource management and the powers they embed in addressing pre-existing gender inequalities. More broadly, mainstreaming gender and a human rights perspective into policy processes would also require the disruption of male patronage networks, the challenging of patriarchal gender norms, and stronger efforts to ensure the meaningful inclusion of all rights holders, particularly diverse groups of women and indigenous communities, in policy development and decision-making.

Notes

1 Parts of this conclusion have drawn inspiration from the DEMETER Research Brief 2/2020 on *Agricultural and Land Commercialisation: Do They Foster Gender Equality and The Right to Food?* https://r4d-demeter.info/wp-content/uploads/2020/11/Research_Brief_2_2020.pdf.

2 As our chapters on Ghana in this book focused on smallholder commercial agriculture, the labour questions of these two types of agricultural commercialization did not receive much attention.

References

Atupare, P. (2016) Qualitative Research Report: East Gonja District, Ghana. DEMETER Working Paper. Accra: Institute of Statistical Social and Economic Research (ISSER), University of Ghana.

Beban, A. and Bourke Martignoni, J. (2021) '"Now the Forest Is Over": Transforming the Commons and Remaking Gender in Cambodia's Uplands', *Frontiers in Sustainable Food Systems* 5, 1–15.

Bourke Martignoni, J. (2018) 'Engendering the Right to Food? International Human Rights Law, Food Security and the Rural Woman', *Transnational Legal Theory* 9(3–4), 400–414. https://doi.org/10.1080/20414005.2018.1568789.

Bourke Martignoni, J. (2020) 'The Right to Food', in Dugard, J., Porter, B., Ikawa, D. and Chenwi, L. (eds), *Research Handbook on Economic, Social and Cultural Rights as Human Rights*, London, Edward Elgar, pp. 138–159.

Bourke Martignoni, J. (2021) 'A Feminist Methodology for Implementing the Right to Food in Agrarian Communities: Reflections from Cambodia and Ghana', *The Journal of Peasant Studies*, 1459–1484. https://doi.org/10.1080/03066150.2021.1928642.

Claeys, P. and Edelman, M. (2019) 'The United Nations Declaration on the Rights of Peasants and Other People Working in Rural Areas', *The Journal of Peasant Studies*: 1–68, Online 24 October. https://doi.org/10.1080/03066150.2019.1672665.

FIAN International (2020a) *Monitoring Report on the Right to Food and Nutrition during COVID-19.* https://www.fian.org/files/files/Covid_Monitoring_Report_-Template_EN(1).pdf.

FIAN International (2020b) *Monitoring Toolkit on the Right to Food and Nutrition during COVID-19.* https://www.fian.org/files/files/Monitoring_Toolkit_RtFN_During_Covid-19.pdf.

FIAN International (2020c) *A Recipe for Disaster: COVID Response Based on the Industrial Food System.* Press release, 17 June 2020. https://www.fian.org/en/press-release/article/a-recipe-for-disaster-covid-response-based-on-the-industrial-food-system-2510.

Golay, C. (2019) *The Implementation of the United Nations Declaration on the Rights of Peasants and Other People Working in Rural Areas*, Research Brief, Geneva Academy of International Humanitarian Law and Human Rights.

Golay, C. (2020) *The Right to Land and UNDROP. How Can We Use the United Nations Declaration on the Rights of Peasants and Other People Working in Rural Areas to Protect the Right to Land?*, International Land Coalition, Geneva Academy of International Humanitarian Law and Human Rights.

Graziano Da Silva, J. (2019) *From Fome Zero to Zero Hunger – A Global Perspective*, FAO, Rome, http://www.fao.org/3/ca5524en/CA5524EN.pdf.

Joshi, S. (2020a) 'Working Wives: Gender, Labour and Land Commercialisation in Ratanakiri, Cambodia', *Globalizations* 17(1): 1–15.

Joshi, S. (2020b) 'Contesting Land Grabs, Negotiating Statehood: The Politics of International Accountability Mechanisms and Land Disputes in Rural Cambodia', *Third World Quarterly* 41(9): 1615–1633.

Lanz, K., Prügl, E. and Gerber, J.-D. (2020) 'The Poverty of Neoliberalized Feminism: Gender Equality in a "Best Practice" Large-Scale Land Investment in Ghana.' *The Journal of Peasant Studies* 47(3): 525–543. https://doi.org/10.1080/03066150.2019.1602525.

Merry, S. and Levitt, P. (2017) 'The Vernacularization of Women's Human Rights', in Hopgood, S., Snyder, J. and Vinjamuri, L. (eds), *Human Rights Futures*, Cambridge, Cambridge University Press, pp. 213–236. https://doi.org/10.1017/9781108147767.009

Mawunyo Dzanku, F., Tsikata, D. and Adu Ankrah, D. (2021) 'The Gender and Geography of Agricultural Commercialisation: What Implications for the Food

Security of Ghana's Smallholder Farmers?' *The Journal of Peasant Studies*, 48, no. 7: 1507–1536, https://doi.org/ 10.1080/03066150.2021.1945584

Tsikata, D. and Yaro, J.A. (2014) 'When a Good Business Model Is Not Enough: Land Transactions and Gendered Livelihood Prospects in Rural Ghana.' *Feminist Economics* 20(1): 202–226. https://doi.org/10.1080/13545701.2013.866261.

Index

Note: **Bold** page numbers refer to tables; *italic* page numbers refer to figures and page numbers followed by "n" denote endnotes.

Kolibri 360 GmbH
Ferdinand-Jühlke-Straße 7
99095 Erfurt, Deutschland
produktsicherheit@kolibri360.de